Exam Success
IEE Code of Practice 2377
Mark Coles and Jonathan Elliott

First published 2007
Reprinted 2008

© 2007 The City and Guilds of London Institute
All direct quotation from the IEE Code of
Practice and answers © 2007 The Institution of
Engineering and Technology

City & Guilds is a trademark of the
City and Guilds of London Institute
The IEE is an imprint of the Institution
of Engineering and Technology

ISBN-13: 978 0 86341 805 1

Cover and book design by CDT Design Ltd
Implementation by SMITH, Karl Shanahan
Typeset in Congress Sans and Gotham
Printed in the UK by Burlington Press

Exam Success
IEE Code of Practice 2377
Mark Coles and Jonathan Elliott

City & Guilds Level 3 Certificate in the Code of Practice for In-service Inspection and Testing of Electrical Equipment (2377)

City & Guilds is the UK's leading provider of vocational qualifications, offering over 500 awards across a wide range of industries, and progressing from entry level to the highest levels of professional achievement. With over 8500 centres in 100 countries, City & Guilds is recognised by employers worldwide for providing qualifications that offer proof of the skills they need to get the job done.

Copies may be obtained from:
Teaching & Learning Materials
City & Guilds
1 Giltspur Street
London EC1A 9DD
For publications enquiries:
T +44 (0)20 7294 4113
F +44 (0)20 7294 3414
E-mail learningmaterials@cityandguilds.com

The Institution of Engineering and Technology is the new institution formed by the joining together of the IEE (The Institution of Electrical Engineers) and the IIE (The Institution of Incorporated Engineers). The new institution is the inheritor of the IEE brand and all its products and services including the IEE Wiring Regulations (BS 7671), the IEE Code of Practice and supporting material.

Copies may be obtained from:
The Institution of Engineering and Technology
P.O. Box 96
Stevenage
SG1 2SD, UK
T +44 (0)1438 767 328
E-mail sales@theiet.org
www.theiet.org

Contents

Introduction   6

The exam

The exam   10

Sitting a City & Guilds online examination   12

Frequently asked questions   18

Exam content   20

Tips from the examiner   26

Exam practice 1 : Management (2377–100)

Sample test   30

Questions and answers   40

Answer key   62

Exam practice 2 : Inspection and testing (2377–200)

Sample test 1   64

Questions and answers   71

Answer key   86

Exam practice 3 : Inspection and testing (2377–200)

Sample test 2   88

Questions and answers   95

Answer key   112

More information

Further reading   114

Online resources   115

Further courses   116

Introduction

How to use this book

This book has been written as a study aid for the City & Guilds Level 3 Certificate in the Code of Practice for In-service Inspection and Testing of Electrical Equipment (2377). It sets out methods of studying, offers advice on exam preparation and provides details of the scope and structure of the examination, alongside sample questions with fully worked-through answers. Used as a study guide for exam preparation and practice, it will help you to reinforce and test your existing knowledge, and will give you guidelines and advice about sitting the exam. You should try to answer the sample test questions under exam conditions (or as close as you can get) and then review all of your answers. This will help you to become familiar with the types of question that might be asked in the exam and also give you an idea of how to pace yourself in order to complete all questions comfortably within the time limit. This book cannot guarantee a positive exam result, but it can play an important role in your overall revision programme, enabling you to focus your preparation and approach the exam with confidence.

IEE Code of Practice for In-service Inspection and Testing of Electrical Equipment (Third Edition)

You will need a copy of the *IEE Code of Practice for In-service Inspection and Testing of Electrical Equipment* to be able to answer the sample questions and in order to revise for the examination. The *IEE Code of Practice* is widely regarded as the standard code of practice for the electrical industry in respect of safe and effective inspection and testing of those items of electrical equipment that do not form a part of the fixed installation.

City & Guilds Level 3 Certificate in the Code of Practice for In-service Inspection and Testing of Electrical Equipment (2377)

There are two exam options and you may choose to take one or both:

Management of Electrical Equipment Maintenance Certificate (2377–100)
If you are a manager having responsibility for maintenance within a building, or for an organization responsible for inspection and testing, you will need to be aware of your responsibilities under the Electricity at Work Regulations 1989. You need to be able to interpret the relevant legislation and assess the risks in respect of the electrical equipment within your charge, or which you are contracted to inspect and test. As a manager you are also responsible for the maintenance of records relating to the inspection and testing of appliances and equipment and to arrange for the re-inspection and re-testing of such items at appropriate intervals. You must therefore be able to interpret

the results of the inspection and testing process and act appropriately using the information obtained. The syllabus of the Management of Electrical Equipment Maintenance Certificate covers mainly sections 1 to 8 of the *IEE Code of Practice*. Successful completion of this Level 3 certificate will show you have a sound working knowledge of the content and application of these parts of the Code of Practice.

Inspection and Testing of Electrical Equipment Certificate (2377–200)

If you are carrying out inspection and testing on items of electrical equipment you must be competent to undertake the inspection and, as appropriate, testing of electrical equipment and appliances, having due regard for the safety of yourself and others who may be affected. As such, you must be aware of the typical causes of damage which may occur to electrical equipment and appliances and the flexible cables while such items are in use. You need to be able to identify items being inspected/tested in terms of their equipment class and type, in order to use appropriate testing methods to enable meaningful test results to be obtained without causing damage to the equipment or appliance. You must also be able to record the results of the inspection/testing process on appropriate forms. You must be able to report such results to managers responsible for the maintenance of electrical equipment/appliances and inform the users of equipment whether items are suitable for continued use via suitable labelling.

Successful completion of this Level 3 certificate will show you have a sound working knowledge of the content and application of those parts of the Code of Practice that relate to equipment construction, inspection, combined inspection/testing, and recording of test results.

Finding a centre

In order to take one or both of the exams, you must register at an approved City & Guilds centre. You can find your nearest centre by looking up the qualification number 2377 on www.cityandguilds.com. The IET is an accredited centre and runs online exams in different parts of the country. For more details, see www.theiet.org.

At each centre, the Local Examinations Secretary will enter you for the award, collect your fees, arrange for your assessment to take place and correspond with City & Guilds on your behalf. The Local Examinations Secretary also receives all of your certificates and correspondence from City & Guilds. Most centres will require you to attend a course of learning before entering you for the examination. These are typically one-day intensive courses per certificate and courses running across a number of weeks, once a week for two to three hours.

Awarding and reporting

When you complete a City & Guilds 2377 Certificate online examination, you will be given your provisional results, as well as a breakdown of your individual performance in the various areas of the examination. This is a useful diagnostic tool if you fail the exam, as it enables you to identify your individual strengths and weaknesses across the different topics.

A Certificate is issued automatically when you have been successful in the assessment but it will not indicate a grade or percentage pass. Your centre will receive your Notification of Candidate's Results and Certificate. Any correspondence is conducted through the centre. The centre will also receive consolidated results lists detailing the performance of all candidates entered.

If you have particular requirements that will affect your ability to attend and take the examination, then your centre should refer to City & Guilds policy document 'Access to Assessment: Candidates with Particular Requirements'.

The exam

The exam	10
Sitting a City & Guilds online examination	12
Frequently asked questions	18
Exam content	20
Tips from the examiner	26

The exam

The exam

There are two examinations relating to the *IEE Code of Practice*, both having a multiple-choice format. You can sit either paper on its own, or you can sit both.

The Management of Electrical Equipment Maintenance exam consists of 45 questions, which you will have one and a half hours to answer. The Inspection and Testing of Electrical Equipment exam consists of 30 questions, which you will have one hour to answer. Each is separately certificated. Both tests are offered on GOLA, a simple online service that does not require strong IT skills. GOLA uses a bank of questions set and approved by City & Guilds. Each candidate receives randomized questions, so no two candidates will sit exactly the same test.

Both exams always follow a set structure, based on the content of the *IEE Code of Practice* broken down into sections. The tables below outline the parts of the exams and the number of questions in each part. They also shows the weighting – so you can see how important each part is in determining your final score.

Paper no: 2377–100
Paper title: Management of Electrical Equipment Maintenance
Duration: 1½ hours   **No. of questions:** 45

Outcome	Topic	Weighting %	No. of questions
1.1	Law and scope of legislation relevant to the management of electrical equipment maintenance	16	7
1.2	Types, use and testing of electrical equipment used for in-service inspection and testing	22	10
1.3	Categories, frequency and practicalities of in-service inspection and testing	27	12
1.4	Procedures, documentation and user responsibilities that are required for in-service inspection and testing	22	10
1.5	Training that is required for in-service inspection and testing	5	2
1.6	Appropriate test instruments and how they are used within in-service inspection and testing	8	4
	Total	100	45

Paper no: 2377–200
Paper title: Inspection and Testing of Electrical Equipment
Duration: 1 hour   **No. of questions:** 30

Outcome	Topic	Weighting %	No. of questions
2.1	Equipment construction	20	6
2.2	Inspection	20	6
2.3	Combined inspection and testing	20	6
2.4	The use of instruments and recording of data	20	6
2.5	Equipment	20	6
	Total	100	30

Notes

Notes

Sitting a City & Guilds online examination

The test will be taken under usual exam conditions. You will not be able to refer to any materials or publications other than the one that is approved for this test (the *IEE Code of Practice for In-service Inspection and Testing of Electrical Equipment*). You will not be allowed to take your mobile phone into the exam room and you cannot leave the exam room unless you are accompanied by one of the test invigilators. If you leave the exam room unaccompanied before the end of the test period, you may not be allowed to come back into the exam.

When you take a City & Guilds test online, you can go through a tutorial to familiarise yourself with the online procedures. When you are logged on to take the exam, the first screen will give you the chance to go into a tutorial. The tutorial shows how the exam will be presented and how to get help, how to move between different screens, and how to mark questions that you want to return to later.

Please work through the tutorial before you start your examination.

This will show you how to answer questions and use the menu options to help you complete the examination.

Please note that examination conditions now apply.

The time allowed for the tutorial is 10 minutes.

Click on Continue to start the tutorial or Skip to go straight to the examination.

| Skip | Continue |

The sample questions in the tutorial are unrelated to the exam you are taking. The tutorial will take 10 minutes, and is not included in the test time. The test will only start once you have completed or skipped the tutorial. A screen will appear that gives the exam information (the time, number of questions and name of the exam).

City& Guilds

Examination: 2377–200 Inspection and Testing of Electrical Equipment

Number of questions: 30

Time allowed: 60 minutes

Note: Examination conditions now apply.

The next screen that will appear is the Help screen, which will give you instructions on how to navigate through this examination. Please click OK to view the Help screen.

The time allowed for the examination will start after you have left the Help screen.

A warning message will appear 5 minutes before the end of the examination.

OK

After clicking 'OK', the help screen will appear. Clicking the 'Help' button on the tool bar at any time during the exam will recall this screen.

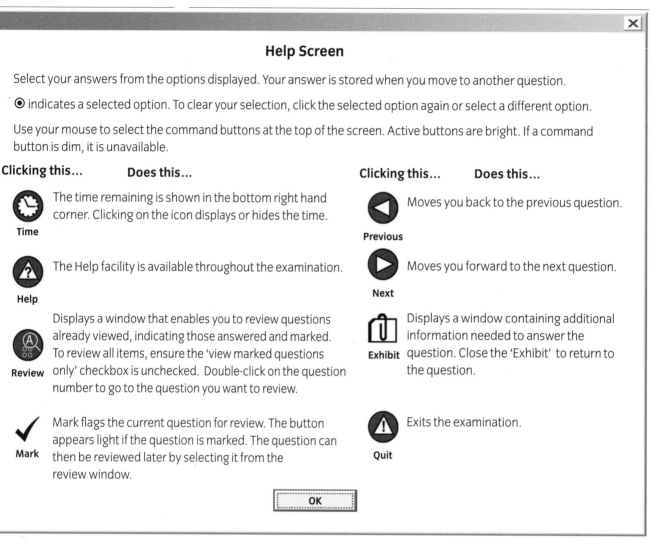

Help Screen

Select your answers from the options displayed. Your answer is stored when you move to another question.

◉ indicates a selected option. To clear your selection, click the selected option again or select a different option.

Use your mouse to select the command buttons at the top of the screen. Active buttons are bright. If a command button is dim, it is unavailable.

Clicking this…	Does this…	Clicking this…	Does this…
Time	The time remaining is shown in the bottom right hand corner. Clicking on the icon displays or hides the time.	**Previous**	Moves you back to the previous question.
Help	The Help facility is available throughout the examination.	**Next**	Moves you forward to the next question.
Review	Displays a window that enables you to review questions already viewed, indicating those answered and marked. To review all items, ensure the 'view marked questions only' checkbox is unchecked. Double-click on the question number to go to the question you want to review.	**Exhibit**	Displays a window containing additional information needed to answer the question. Close the 'Exhibit' to return to the question.
Mark	Mark flags the current question for review. The button appears light if the question is marked. The question can then be reviewed later by selecting it from the review window.	**Quit**	Exits the examination.

OK

After clicking 'OK' while in the help screen, the exam timer will start and you will see the first question. The question number is always shown in the lower left-hand corner of the screen. If you answer a question but wish to return to it later, then you can click the 'Flag' button. When you get to the end of the test, you can choose to review these flagged questions.

Notes

An electrical heater has a mass of 16 kg and has castors to facilitate movement. The equipment type is

○   a   fixed

○   b   transferable

○   c   movable

○   d   hand-held.

Question Number 1

If you select 'Quit' on the tool bar at any point, you will be given the choice of ending the test. **If you select 'Yes', you will not be able to go back to your test.**

If you click 'Time' on the tool bar at any point, the time that you have left will appear in the bottom right-hand corner. When the exam timer counts down to five minutes remaining, a warning will flash on to the screen.

Some of the questions in the test may be accompanied by pictures. The question will tell you whether you will need to click on the 'Exhibit' button to view an image.

When you reach the final question and click 'Next', you will reach a screen that allows you to 'Review your answers' or 'Continue' to end the test. You can review all of your answers or only the ones you have flagged. To review all your answers, make sure that the 'view marked questions only' checkbox is unchecked (click to uncheck). After you have completed your review, you can click 'Continue' to end the test.

City& Guilds

You have answered   30   questions out of a total of   30

To check your answers and return to the examination, click on the Review button. If your time has expired, you cannot return to the examination.

If you wish to submit your answers and end the examination, please click the Continue button.

Clicking Continue will end the examination.

| Review | Continue |

Once you choose to end the exam by clicking 'Continue', the 'Test completed' screen will appear. Click on 'OK' to end the exam.

At the end of the exam, you will be given an 'Examination Score Report'. This gives a provisional grade (pass or fail) and breakdown of your score by section. This shows your performance in a bar chart and in percentage terms, which allows you to assess your own strengths and weaknesses. If you did not pass, it gives valuable feedback on which areas of the course you should revise before resitting the exam.

Notes

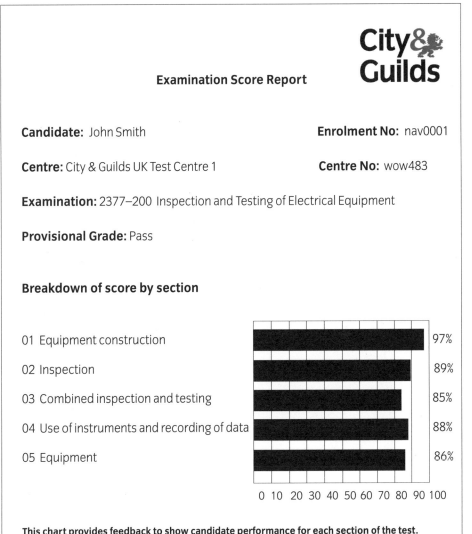

Examination Score Report

City& Guilds

Candidate: John Smith

Enrolment No: nav0001

Centre: City & Guilds UK Test Centre 1

Centre No: wow483

Examination: 2377–200 Inspection and Testing of Electrical Equipment

Provisional Grade: Pass

Breakdown of score by section

01 Equipment construction — 97%
02 Inspection — 89%
03 Combined inspection and testing — 85%
04 Use of instruments and recording of data — 88%
05 Equipment — 86%

0 10 20 30 40 50 60 70 80 90 100

This chart provides feedback to show candidate performance for each section of the test.
It should be used along with the Test Specification, which can be found in the Scheme Handbook.

Frequently asked questions

When can I sit the paper?
You can sit either exam at any time, as there are no set exam dates. You may need to check with your centre when it is able to hold exam sessions.

Can I use reference books in the test?
Yes, for both exams you can use a copy of the IEE Code of Practice for In-service Inspection and Testing of Electrical Equipment (Third Edition).

How many different parts of the test are there?
For the Management of Electrical Equipment Maintenance exam there are 45 questions, which cover six different parts of the syllabus. For the Inspection and Testing of Electrical Equipment exam there are 30 questions, which cover five different parts of the syllabus.

Do I have a time limit for taking the test?
You have one and a half hours to complete the Management exam and one hour to complete the Inspection and Testing exam.

Do I need to be good at IT to do the test online?
No, the system is really easy to use, and you can practise before doing the test. There is also a practice GOLA test available to try on the IEE website.

What happens if the computer crashes in the middle of my test?
This is unlikely, because of the way the system has been designed. If there is some kind of power or system failure, then your answers will be saved and you can continue on another machine if necessary.

Can people hack into the system and cheat?
There are lots of levels of security built into the system to ensure its safety. Also, each person gets a different set of questions, which makes it very difficult to cheat.

Can I change my answer?
Yes, you can change your answers quickly, easily and clearly at any time in the test up to the point where you end the exam. With any answers you feel less confident about, you can click the 'Flag' button, which means you can review these questions before you end the test.

How do I know how long I've got left to complete the test?
You can check the time remaining at any point during the exam by clicking on the 'Clock' icon in the tool bar. The time remaining will come up on the bottom right corner of the screen.

Is there only one correct A, B, C or D answer to multiple-choice questions?
Yes.

What happens if I don't answer all of the questions?
You should attempt to answer all of the questions. If you find a question difficult, mark it using the 'Flag' button and return to it later.

What grades of pass are there?
A Pass or a Fail.

When can I resit the test if I fail?
You can resit the exam at any time and as soon as you and your tutor decide it is right for you, subject to the availability of the online examination.

Notes

Exam content

To help you to fully understand the exam content, this chapter looks at the requirements of the two exams by breaking down each section and referring to the relevant parts of the *IEE Code of Practice for In-service Inspection and Testing of Electrical Equipment*. As you work through, remember that the Code of Practice has a useful index. If you don't know where to look to find the relevant information, look in the index.

2377–100 Management of Electrical Equipment Maintenance

For the most part, the answers for this exam can be found in Part 1 of the Code of Practice (Administration of inspection and testing) so you should look for the answers in this part first. However, it will also be necessary to refer to Part 2 (Inspection and testing) or Part 3 (Appendices) to answer some questions.

Outcome 1.1

Law and scope of legislation relevant to the management of electrical equipment maintenance

There will be 7 questions in this section, which accounts for 16 per cent of the maximum achievable mark.

To show that you are conversant with the Code of Practice, you are required to be able to:

1.1.1 State the related requirements of:
   a   the Health and Safety at Work etc Act, ie duties of employers, employees, visitors to premises and the self employed
   b   the Management of Health and Safety at Work Regulations, Regulation 3(1) – Risk Assessment
   c   the Provision and Use of Work Equipment Regulations
   d   the Electricity at Work Regulations
   e   the Workplace (Health, Safety & Welfare) Regulations
1.1.2 Identify the scope of legislation with regard to the system voltage
1.1.3 Identify those properties and premises to which the Act and Regulations apply and those which are excluded
1.1.4 State the guidance given by the Health and Safety Executive relating to electrical equipment (in particular, the guidance on procedures for isolation of supplies)
1.1.5 State the legal requirements to maintain electrical equipment in a safe condition in work premises

1.1.6 State the importance of, and reasons for, inspecting and testing electrical equipment and systems.

Outcome 1.2

Types, use and testing of electrical equipment used for in-service inspection and testing

There will be 10 questions in this section, which accounts for 22 per cent of the maximum achievable mark.

To show that you are conversant with the Code of Practice, you are required to be able to:

1.2.1 Classify types and construction of electrical equipment identified in the Code of Practice

1.2.2 State how the construction of electrical equipment provides protection against electric shock

1.2.3 State the specific requirements for inspection and testing of extension leads

1.2.4 State the different types of tests that may be utilised during the life of equipment

1.2.5 State what tests may be required following repairs to electrical equipment.

Outcome 1.3

Categories, frequency and practicalities of in-service inspection and testing

There will be 12 questions in this section, which accounts for 27 per cent of the maximum achievable mark.

To show that you are conversant with the Code of Practice, you are required to be able to:

1.3.1 Identify the categories of inspection and testing

1.3.2 State the factors governing the frequency of in-service inspection and testing

1.3.3 State the purpose of Table 7.1 *Initial frequency of inspection and testing of equipment* (see Section 7 of the Code of Practice)

1.3.4 Detail the visual inspection/examination requirements and note the contents of Appendix VII

1.3.5 Detail the tests that are required for in-service inspection and testing of equipment.

Notes

Outcome 1.4

Procedures, documentation and user responsibilities that are required for in-service inspection and testing

There will be 10 questions in this section, which accounts for 22 per cent of the maximum achievable mark.

To show that you are conversant with the Code of Practice, you are required to be able to:

1.4.1 State the need for the model forms in the Code of Practice to provide data on the results of electrical equipment maintenance

1.4.2 State the need for equipment identity, numbering and labelling

1.4.3 Record information, testing requirements and results

1.4.4 Interpret test results

1.4.5 Identify instrument requirements

1.4.6 State the procedure for dealing with equipment found to be faulty.

Outcome 1.5

Training that is required for in-service inspection and testing

There will be 2 questions in this section, which accounts for 5 per cent of the maximum achievable mark.

To show that you are conversant with the Code of Practice, you are required to be able to:

1.5.1 State the Electricity at Work Regulations' requirements for maintaining electrical equipment, including the need for competence when managing such maintenance

1.5.2 State the training requirements for users of equipment with regard to:
   a   the safe use of equipment
   b   identifying damaged equipment and flexible cables and cords, together with connecting plugs and couplers, etc

1.5.3 State the training requirements for managers to:
   a   undertake risk assessment in the workplace
   b   maintain records of electrical maintenance
   c   interpret the results of electrical tests

1.5.4 State the training requirements for inspectors.

Outcome 1.6

Appropriate test instruments and how they are used within in-service inspection and testing

There will be 4 questions in this section, which accounts for 8 per cent of the maximum achievable mark.

To show that you are conversant with the Code of Practice, you are required to be able to:

1.6.1 List instruments that are suitable for testing electrical equipment, together with means of on-going checks of instrument accuracy

1.6.2 State types of continuity tester, indicating in each case the instrument's short circuit test current(s)

1.6.3 State means of determining equipment insulation resistance, earth leakage/touch current measurement, indicating instrument voltage and current

1.6.4 State procedures for conducting electrical tests.

2377–200 Inspection and Testing of Electrical Equipment

For the most part, the answers for this exam can be found in Part 2 of the Code of Practice (Inspection and testing) so you should look for the answers in this part first. However, it will also be necessary to refer to Part 1 (Administration of inspection and testing) or Part 3 (Appendices) to answer some questions.

Outcome 2.1

Equipment construction
There will be 6 questions in this section, which accounts for 20 per cent of the maximum achievable mark.

To show that you are conversant with the Code of Practice, you are required to be able to:

2.1.1 State the different types of equipment, together with their form of construction and classifcation marks

2.1.2 Identify how electric shock can occur through lack of basic protection or inadequate fault protection

2.1.3 State how equipment construction provides protection against electric shock

2.1.4 State the effect of equipment flexible leads and cord sets' resistance on disconnection times under earth fault conditions

2.1.5 State the need for RCD protection for certain items of equipment.

Notes

Outcome 2.2

Inspection

There will be 6 questions in this section, which accounts for 20 per cent of the maximum achievable mark.

To show that you are conversant with the Code of Practice, you are required to be able to:

2.2.1 State essential initial frequencies for:
- a   user checks
- b   formal visual inspection/examination
- c   combined inspection/examination and testing

2.2.2 Identify items that the user should be competent to inspect

2.2.3 Describe what needs to be considered when carrying out a formal visual inspection with regard to:
- a   suitability of equipment for the environment
- b   good housekeeping
- c   suitability of equipment for intended use
- d   requirements for switching of equipment and disconnection of supplies
- e   condition of equipment and connecting cable or cord.

Outcome 2.3

Combined inspection and testing

There will be 6 questions in this section, which accounts for 20 per cent of the maximum achievable mark.

To show that you are conversant with the Code of Practice, you are required to be able to:

2.3.1 List the factors to be taken into account when conducting visual inspection/examination of equipment

2.3.2 State the purpose of the following tests:
- a   earth continuity (earth bonding), including circumstances which would require the use of a purpose-built continuity tester
- b   insulation resistance and earth leakage/touch current measurement, including circumstances which would require the use of a purpose-built insulation resistance tester
- c   substitute/alternative leakage measurement
- d   operation/load checking
- e   polarity

f   where an extension lead incorporates an RCD, an electrical test should be carried out to confirm correct operation of the device

g   microwave ovens should be checked for leakage levels at appropriate intervals

2.3.3   State which tests should not be carried out

2.3.4   Describe the various types of instruments that may be used for testing electrical equipment.

Outcome 2.4

The use of instruments and recording of data

There will be 6 questions in this section, which accounts for 20 per cent of the maximum achievable mark.

To show that you are conversant with the Code of Practice, you are required to be able to:

2.4.1   State minimum and maximum acceptable conductor and insulation resistances, together with 'leakage' and 'touch' currents

2.4.2   State the information to be recorded on an identification label

2.4.3   Recognise deteriorating equipment by interpretation of test results.

Outcome 2.5

Equipment

There will be 6 questions in this section, which accounts for 20 per cent of the maximum achievable mark.

To show that you are conversant with the Code of Practice, you are required to be able to:

2.5.1   State the requirements for procedures to deal with damaged and faulty equipment

2.5.2   State particular considerations for appliance couplers and cord sets

2.5.3   State recommended testing procedures for microwave ovens

2.5.4   Identify particular considerations for information technology (IT) equipment having high protective conductor currents and associated testing limitations

2.5.5   State requirements for appliance flexible cables and cords and their overcurrent protection by the two standard sizes of plug fuse

2.5.6   State tests that may be required following equipment repair.

Tips from the examiner

The following tips are intended to aid confidence and test performance. Some are more general and would apply to most exams. Others are more specific, either because of the format of this test (multiple choice) or the nature of the subject.

✔ If you rarely use a computer, try to get some practice beforehand. You need to be able to use a mouse to move a cursor arrow around a computer screen, as you will use the cursor to click on the correct answer in the exam.

✔ Take the time to familiarise yourself with the structure and content of the *IEE Code of Practice* prior to the exam.

✔ Make the most of the course you will attend before taking the test. Try to attend all sessions and be prepared to devote time outside the class to revise for the exam.

✔ On the day of the exam, allow plenty of time for travel to the centre and arrive at the place of the exam at least 10 minutes before it's due to start so that you have time to relax and get into the right frame of mind.

✔ Listen carefully to the instructions given by the invigilator.

✔ Read the question and every answer before making your selection. Do not rush – there should be plenty of time to answer all the questions.

✔ Look at the exhibits where instructed. Remember, an exhibit supplies you with information that is required to answer the question.

✔ Attempt to answer all the questions. If a question is not answered, it is marked as wrong. Making an educated guess improves your chances of choosing the correct answer. Remember, if you don't select an answer, you will definitely get no marks for that question.

✔ Don't worry about answering the questions in the order in which they appear in the exam. Choose the 'Flag' option on the tool bar to annotate the questions you want to come back to. If you spend too much time on questions early on, you may not have time to answer the later questions, even though you know the answers.

✔ Although not absolutely necessary, some candidates find it useful to bring a basic, non-programmable calculator to the exam.

✔ It is not recommended that you memorise any of the material presented here in the hope it will come up in the exam. The exam questions featured in this book will help you to gauge the kinds of questions that might be asked. It is highly unlikely you will be asked any identical questions in the exam but you may see variations on certain themes.

Notes

Notes

Exam practice 1: Management (2377–100)

Sample test	30
Questions and answers	40
Answer key	62

Exam practice 1

Notes

Sample test

The sample test below has 45 questions, which is the same number as the online Management of Electrical Equipment Maintenance exam (2377–100), and its structure follows that of the online exam. The test appears firstly without answers, so you can use it as a mock exam. It is then repeated with answers and explanations. Finally, there is an answer key for easy reference.

Answer the questions by filling in the circle next to your chosen option.

Outcome 1.1 The law and scope of legislation

1 **Which of the following is a non-statutory document designed to assist with portable appliance testing?**

○ a   Electricity at Work Regulations 1989
○ b   Provision and Use of Work Equipment Regulations 1998 (PUWER)
○ c   *IEE Code of Practice for In-service Inspection and Testing of Electrical Equipment*
○ d   *IEE Guidance Note 3: Inspection and Testing*

2 **The Provision and Use of Work Equipment Regulations 1998 states that equipment must be in an efficient state. The term 'efficient' relates to equipment**

○ a   productivity
○ b   safety
○ c   cost
○ d   power consumption.

3 **The Provision and Use of Work Equipment Regulations 1998 requires equipment to be maintained in good order. This duty is placed upon the**

○ a   employee
○ b   contractor
○ c   user
○ d   employer.

4 Which of the following would <u>not</u> be covered under the scope of the Electricity at Work Regulations 1989?

Notes

- ○ a   A computer monitor in an office
- ○ b   A bench grinder in a domestic garden shed
- ○ c   A 9 V d.c. calculator used in a shop
- ○ d   A hand-held two-way radio charger in a security kiosk

5 Class I appliances depend on the integrity of a building's fixed electrical installation. Guidance on the inspection and testing of a fixed electrical installation can be found in

- ○ a   *IEE Guidance Note 3: Inspection and Testing*
- ○ b   HSE Guidance Note GS38
- ○ c   *BS 7671: Requirements for Electrical Installations*
- ○ d   HSE Guidance Note PM29.

6 Specific guidance on maintaining portable and transportable equipment may be found in HSE document

- ○ a   GS38
- ○ b   HS(G)107
- ○ c   GS6
- ○ d   HS(G)47.

7 The Electricity at Work Regulations 1989 does <u>not</u> apply to electrical equipment

- ○ a   used at work but owned by an employee
- ○ b   powered by a battery
- ○ c   sold by an employer and used only at home
- ○ d   used at home for work purposes.

Outcome 1.2 Electrical equipment

8 A free-standing 3 kW oil-filled radiator with a mass of 17 kg is considered to be

- ○ a   portable equipment
- ○ b   stationary equipment
- ○ c   fixed equipment
- ○ d   movable equipment.

9 **A panel heater is secured to a wall and supplied by a standard 13 A plug. The equipment type is**

- ○ a   portable
- ○ b   hand-held
- ○ c   fixed
- ○ d   transportable.

10 **A 12 V compressor relies on the supply voltage as a means of electric shock protection. This equipment is classified as**

- ○ a   Class I
- ○ b   Class II
- ○ c   Class III
- ○ d   Class IV.

11 **To provide protection against electric shock, Class II equipment is**

- ○ a   reliant on a SELV source of supply
- ○ b   not reliant on the fixed electrical installation
- ○ c   reliant on the fixed electrical installation including a reliable earth
- ○ d   not reliant on basic insulation and additional safety precautions.

12 **Extension leads with a length of 30 m should**

- ○ a   be protected by a fuse to BS 1362
- ○ b   have a cross-sectional area of 1.5 mm^2
- ○ c   be protected by a suitably rated RCD
- ○ d   have a cross-sectional area of 1.25 mm^2.

13 **Following major repair to an electrical appliance, it is recommended that the appliance is subjected to**

- ○ a   a user inspection
- ○ b   a consumer inspection
- ○ c   production testing
- ○ d   type testing to an appropriate standard.

14 **Class III equipment relies on electric shock protection by means of a SELV supply. A SELV supply shall <u>not</u> exceed**

- ○ a   12 V a.c.
- ○ b   24 V a.c.
- ○ c   50 V a.c.
- ○ d   120 V a.c.

15 A two-core extension lead <u>must</u> be fitted with

- ○ a   an RCD
- ○ b   a three-pin socket
- ○ c   a two-pin socket
- ○ d   a BS EN 60309 plug.

16 An item of equipment subjected to type testing will usually be

- ○ a   destroyed
- ○ b   sold as seen
- ○ c   used for spares
- ○ d   put into service.

17 A metal-cased item of equipment is supplied using a two-core cord set and has a classification mark as below. This item is

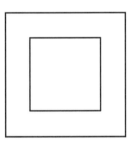

- ○ a   dangerous and should be removed from service
- ○ b   Class I
- ○ c   dangerous and should be fitted with a three-core cord
- ○ d   Class II.

Outcome 1.3 In-service inspection and testing

18 A category of inspection and testing that may <u>not</u> require records to be kept is the

- ○ a   user check
- ○ b   production test
- ○ c   formal visual inspection
- ○ d   combined inspection and test.

19 The <u>most</u> important check to be carried out on any item of equipment in the workplace is the

- ○ a   in-service test
- ○ b   production test
- ○ c   formal visual inspection
- ○ d   combined inspection and test.

20 Recorded tests may be omitted for Class II equipment

- ○ a   located in an office
- ○ b   supplied using a three-core cord
- ○ c   located in a factory
- ○ d   supplied using a BS EN 60309 plug.

21 <u>Four</u> factors that should influence the frequency of inspection and testing of appliances are the

- ○ a   environment, users, equipment construction and equipment type
- ○ b   users, equipment construction, equipment type and cost involved
- ○ c   equipment construction, equipment type, cost involved and environment
- ○ d   equipment type, cost involved, environment and users.

22 Table 7.1 in Section 7 of the *IEE Code of Practice for In-service Inspection and Testing of Electrical Equipment* provides guidance on initial frequencies of inspection and testing. The intervals between inspection and testing may be changed depending on

- ○ a   the costs involved to carry out the tests
- ○ b   the availability of the inspector
- ○ c   a pattern of failure or damage being established
- ○ d   a suitable time when the equipment is available.

23 Class I equipment is considered to be at a higher risk of damage or deterioration when

- ○ a   located on a construction site
- ○ b   used in a school
- ○ c   located in a factory
- ○ d   located in a hotel.

24 Where the class of an item of equipment cannot be established, the suggested initial frequency of inspecting and testing the equipment shall be

- ○ a   as Class II
- ○ b   as Class I
- ○ c   extended
- ○ d   ignored.

25 A formal visual inspection of equipment will involve careful scrutiny of the appliance,

- ○ a   the cord set, the suitability of the appliance and the environment
- ○ b   the suitability of the appliance, the environment and the fixed electrical installation
- ○ c   the environment, the fixed electrical installation and the cord set
- ○ d   the fixed electrical installation, the cord set and the suitability of the appliance.

26 Prior to a formal inspection of equipment, it should be established whether the equipment has any faults. The most appropriate person to ask is

- ○ a   a responsible person
- ○ b   a competent person
- ○ c   a regular user
- ○ d   the duty holder.

27 Class I appliances must be tested for

- ○ a   earth continuity, insulation resistance (or touch current) and function
- ○ b   insulation resistance (or touch current), function and polarity
- ○ c   function, polarity and dielectric strength
- ○ d   polarity, dielectric strength and earth continuity.

28 Which test or inspection is not applicable to Class II appliance-couplers?

- ○ a   Insulation resistance
- ○ b   Corrosion
- ○ c   Earth continuity
- ○ d   Mechanical damage

29 Which of the following tests may cause damage to vulnerable electronic equipment?

○ a   Earth continuity
○ b   Touch current measurement
○ c   Polarity
○ d   Insulation resistance

Outcome 1.4 Procedures, documentation, etc

30 The <u>main</u> purpose of an equipment register is to

○ a   assist with stock control
○ b   identify corporate thefts
○ c   record maintenance details
○ d   review the frequency of inspection and testing.

31 Which of the following should appear on the label applied to equipment after testing?

○ a   Re-testing due date
○ b   Previous test result data
○ c   Inspector's company name and logo
○ d   Manufacturer's production test certificate

32 The model form that does <u>not</u> require the equipment serial number to be recorded is the

○ a   equipment register (Form VI.1)
○ b   inspection and test record (Form VI.2)
○ c   repair register (Form VI.4)
○ d   faulty equipment register (Form VI.5).

33 A value of insulation resistance recorded as 1.34 MΩ is equivalent to

○ a   13,400 Ω
○ b   134,000 Ω
○ c   1,340,000 Ω
○ d   13,400,000 Ω.

34 The nominal resistance for the protective conductor for a cord set having a cross-sectional area of 1 mm² and a length of 5 m is

- ○ a 0.0975 Ω
- ○ b 0.975 Ω
- ○ c 9.750 Ω
- ○ d 97.50 Ω.

35 An instrument being used to carry out an earth continuity test should produce a test current of not less than 1.5 times the rating of the fuse and not greater than

- ○ a 100 mA
- ○ b 200 mA
- ○ c 26 A
- ○ d 100 A.

36 If a low resistance ohmmeter is used for earth continuity testing, a suitable test current would be

- ○ a 0.01 mA
- ○ b 10 mA
- ○ c 100 mA
- ○ d 500 mA.

37 A test facility, found on certain appliance test instruments, which is a useful means of determining whether or not a dual element heater is fully functioning, is the

- ○ a dielectric strength test
- ○ b load test
- ○ c earth continuity test
- ○ d insulation resistance test.

38 Any equipment that is faulty due to being unsuitable for the use intended should be

- ○ a repaired
- ○ b tested more frequently
- ○ c replaced with an identical item
- ○ d replaced with a suitable item.

39 An item of equipment is found to be unsafe and is reported to the responsible person by the inspector. The correct immediate action is to

- ○ a   have the item inspected more frequently
- ○ b   remove the equipment from use
- ○ c   fix a failure label to the equipment
- ○ d   ask the user to take care when using the equipment.

Outcome 1.5   Training

40 It is recommended that users of appliances receive suitable training in order to carry out

- ○ a   formal inspections
- ○ b   combined inspections and tests
- ○ c   user checks and reports of any faults found
- ○ d   suitable repairs to faulty equipment.

41 Managers of testing organisations should receive training in order to understand their legal responsibilities as laid down in the

- ○ a   Electricity at Work Regulations 1989
- ○ b   Health and Safety at Work etc. Act 1974
- ○ c   Electricity Safety, Quality and Continuity Regulations 2002
- ○ d   *BS 7671: Requirements for Electrical Installations.*

Outcome 1.6   Test instruments

42 The <u>most</u> suitable instrument for conducting an earth continuity test on an item of equipment that is connected directly to a switched fused connection unit is

- ○ a   a low resistance ohmmeter/insulation resistance tester set to 500 V
- ○ b   a low resistance ohmmeter/insulation resistance tester set to low ohms
- ○ c   a portable appliance test instrument having a 13 A plug and socket facility
- ○ d   an earth loop impedance tester set at 20 Ω.

43 It is recommended that an insulation resistance tester applies a test voltage of

○ a   500 V a.c.
○ b   500 V d.c.
○ c   3750 V a.c.
○ d   3750 V d.c.

44 A suitable alternative test for insulation resistance, where a high voltage may damage equipment, is

○ a   an earth continuity test
○ b   a load test
○ c   a functional test
○ d   a measured touch current test.

45 In order to avoid having to re-test a large number of appliances, it is recommended that test equipment is

○ a   replaced yearly
○ b   regularly checked for accuracy
○ c   the most modern model
○ d   able to perform touch current testing.

Questions and answers

The questions in the sample test are repeated below with worked-through answers and extracts from the *IEE Code of Practice for In-service Inspection and Testing of Electrical Equipment* where appropriate. Where references to sections are made and extracts given, these may be found in the Code of Practice publication.

Outcome 1.1 The law and scope of legislation

1   **Which of the following is a non-statutory document designed to assist with portable appliance testing?**

- ○ a   Electricity at Work Regulations 1989
- ○ b   Provision and Use of Work Equipment Regulations 1998 (PUWER)
- ● c   *IEE Code of Practice for In-service Inspection and Testing of Electrical Equipment*
- ○ d   *IEE Guidance Note 3: Inspection and Testing*

Answer c

The *IEE Code of Practice for In-service Inspection and Testing of Electrical Equipment* is designed to assist operatives with portable appliance testing and is a non-statutory document. See Section 5 of the Code of Practice (Types of electrical equipment).

Option a, the Electricity at Work Regulations 1989, is incorrect as this is a statutory document and, as such, must be adhered to. It imposes duties on persons in respect of systems, electrical equipment and conductors. Option b, the Provision and Use of Work Equipment Regulations 1998, requires that electrical equipment is so constructed as to be suitable for the purpose for which it was provided. Option d, *Guidance Note 3*, published by the IEE, gives guidance and recommendations to assist persons carrying out inspection, testing and certification of fixed electrical installations, as opposed to portable appliances.

2   **The Provision and Use of Work Equipment Regulations 1998 states that equipment must be in an efficient state. The term 'efficient' relates to equipment**

- ○ a   productivity
- ● b   safety
- ○ c   cost
- ○ d   power consumption.

Answer b
The term 'efficient' relates to equipment safety (see Section 3.4, Maintenance). It is not related to the other three options – productivity, cost or power consumption.

From Section 3.4 The Approved Code of Practice & Guidance document to the Provision and Use of Work Equipment Regulations 1998 (L22) states that 'efficient' relates to how the condition of the equipment might affect health and safety; it is not concerned with productivity.

3 **The Provision and Use of Work Equipment Regulations 1998 requires equipment to be maintained in good order. This duty is placed upon the**

- ○ a   employee
- ○ b   contractor
- ○ c   user
- ◉ d   employer.

Answer d
Regulation 4(1) of the Provision and Use of Work Equipment Regulations 1998 states that every **employer** shall ensure that work equipment is constructed or adapted so that it is suitable for the purpose for which it has been provided. See Section 3.3 (Who is responsible?).

However, the employee, contractor and user do all have a duty of care under the Health and Safety at Work etc. Act 1974 in relation to work activities.

From Section 3.3 The Provision and Use of Work Equipment Regulations 1998 requires every employer to ensure that equipment is suitable for the use for which it is provided (Reg 4(1)) and only used for work for which it is suitable (Reg 4(3)). They require every employer to ensure equipment is maintained in good order (Reg 5) and inspected as necessary to ensure it is maintained in a safe condition (Reg 6).

4 **Which of the following would <u>not</u> be covered under the scope of the Electricity at Work Regulations 1989?**

- ○ a   A computer monitor in an office
- ◉ b   A bench grinder in a domestic garden shed
- ○ c   A 9 V d.c. calculator used in a shop
- ○ d   A hand-held two-way radio charger in a security kiosk

Notes

Answer b

The Electricity at Work Regulations 1989 applies to all electrical equipment used in, or associated with, places of **work**. This does not apply in the case of option b, using a bench grinder in a domestic garden shed. See Section 3.2 (Scope of the legislation).

Note 2 in Appendix II of the Code of Practice (Legal references and notes) states that 'electrical equipment' as defined in the Electricity at Work Regulations includes every type of electrical equipment, from a 400 kV overhead line to a battery-powered hand-lamp. It is appropriate for the Regulations to apply even at the very lowest end of the voltage or power spectrum because the Regulations are concerned with, for example, explosion risks, which may be caused by very low levels of energy igniting flammable gases, even though there may be no risk of electric shock or burns. Thus no voltage limits appear in the Regulations. The criterion or application is the test as to whether 'danger' (as defined) may arise.

Examiner's tip: Please read questions carefully as this type of 'negative' item often catches candidates out, who read the question too quickly and fail to notice the '**not**'.

From Appendix II, Note II.2   The purpose of the Electricity at Work Regulations 1989 is to prevent death or injury to anyone from any electrical cause as a result of, or in connection with, work activities.

5   **Class I appliances depend on the integrity of a building's fixed electrical installation. Guidance on the inspection and testing of a fixed electrical installation can be found in**

- ◉ a   *IEE Guidance Note 3: Inspection and Testing*
- ○ b   HSE Guidance Note GS38
- ○ c   *BS 7671: Requirements for Electrical Installations*
- ○ d   HSE Guidance Note PM29.

Answer a

Although option c, *BS 7671: Requirements for Electrical Installations*, states the requirements for inspecting and testing a fixed electrical installation, detailed information and guidance is contained in option a, *IEE Guidance Note 3: Inspection and Testing*. See Section 4 (Fixed electrical installation).

Option b, HSE Guidance Note GS38: Electrical test equipment for use by electricians, relates to test probes and leads, in particular those used to apply or measure voltage over 50 V a.c. and 100 V d.c. See Section 10 (Test instruments).

054863

Option d, HSE Guidance Note PM29, offers guidance on the electrical hazards from steam/water pressure cleaners. See Appendix IV of the Code of Practice, Table IV.4.

6 Specific guidance on maintaining portable and transportable equipment may be found in HSE document

- ○ a   GS38
- ◉ b   HSG 107
- ○ c   GS6
- ○ d   HSG 47.

Answer b

HSG 107 provides specific guidance on maintaining portable and transportable electrical equipment. See Appendix IV, Table IV.3.

Options a, c and d are documents with different uses. Option a, HSE Guidance Note GS38: Electrical test equipment for use by electricians, provides information on test probes and leads. Option c, GS6, provides information on the avoidance of danger from overhead electric lines, while option d, HSG 47, provides information on avoiding danger from underground services.

7 The Electricity at Work Regulations 1989 does not apply to electrical equipment

- ○ a   used at work but owned by an employee
- ○ b   powered by a battery
- ◉ c   sold by an employer and used only at home
- ○ d   used at home for work purposes.

Answer c

The Electricity at Work Regulations 1989 requires precautions to be taken against the risk of death or personal injury from any electrical equipment used in work activities. Option c, electrical equipment sold by an employer for use only at home, is no longer owned by the employer and is no longer part of work activities and so the regulations do not apply. See Section 3 (The law).

The regulations apply to options a and d because the equipment is used for work activities. They also apply to option b because the regulations include every type of electrical equipment from high voltage overhead lines to battery-powered equipment. It should be remembered that no voltage limits appear in the regulations and that even very low energy levels may constitute an ignition hazard in some circumstances.

Outcome 1.2 Electrical equipment

8 **A free-standing 3 kW oil-filled radiator with a mass of 17 kg is considered to be**

- ○ a   portable equipment
- ○ b   stationary equipment
- ○ c   fixed equipment
- ◉ d   movable equipment.

Answer d

Movable equipment is defined in Section 5 (Types of electrical equipment). The definitions for the other options do not fit the description of the radiator.

From Section 5.2

An item of movable equipment is equipment which is either:
- 18 kg or less in mass and not fixed, eg an electric compressor, or
- equipment with wheels, castors or other means to facilitate movement by the operator as required to perform its intended use, eg air-conditioning unit.

9 **A panel heater is secured to a wall and supplied by a standard 13 A plug. The equipment type is**

- ○ a   portable
- ○ b   hand-held
- ◉ c   fixed
- ○ d   transportable.

Answer c

Equipment or appliances fastened to a support, such as the panel heater in the question, are considered to be fixed appliances. See Section 5 (Types of electrical equipment) for definitions of all four options.

From Section 5.5 An item of fixed equipment or a fixed appliance is equipment that is fastened to a support or otherwise secured in a specified location, eg hand drier or bathroom heater.

10 **A 12 V compressor relies on the supply voltage as a means of electric shock protection. This equipment is classified as**

- ○ a   Class I
- ○ b   Class II
- ◉ c   Class III
- ○ d   Class IV.

Answer c

See Section 2 (Definitions) for information about Class I, II and III equipment. There is no such classification as Class IV (option d). Section 11.3 (Class III) further clarifies that SELV sources will not exceed 50 volts.

From Section 2 Class III equipment. Equipment in which protection against electric shock relies on a supply at SELV and in which voltages higher than those of SELV are not generated.

11 To provide protection against electric shock, Class II equipment is

○ a   reliant on a SELV source of supply
◉ b   not reliant on the fixed electrical installation
○ c   reliant on the fixed electrical installation including a reliable earth
○ d   not reliant on basic insulation and additional safety precautions.

Answer b

Class II equipment requires additional safety precautions and is not reliant only on the fixed electrical installation. See Section 2 (Definitions).

From Section 2 Class II equipment. Equipment in which protection against electric shock does not rely on basic insulation only, but in which additional safety precautions such as supplementary insulation are provided, there being no provision for the connection of exposed metalwork of the equipment to a protective conductor and no reliance upon precautions to be taken in the fixed wiring of the installation.

12 Extension leads with a length of 30 m should

○ a   be protected by a fuse to BS 1362
○ b   have a cross-sectional area of 1.5 mm^2
◉ c   be protected by a suitably rated RCD
○ d   have a cross-sectional area of 1.25 mm^2.

Answer c

An extension lead exceeding 25 metres should be protected by an RCD with a rated residual operating current not exceeding 30 mA. See Section 15.10.1 (Extension leads).

Limiting the length of extension leads is done predominantly to prevent the effects of voltage drop adversely affecting the performance of connected items. It is also good practice to keep extension leads as short as possible and to route them carefully to minimise the risk of a trip hazard.

Table 15.4 Maximum recommended lengths for extension leads

Cross-sectional area of the core	Maximum length
1.25 mm^2	12 metres
1.5 mm^2	15 metres
2.5 mm^2	25 metres

Any extension lead exceeding the above lengths should be fitted with an RCD with a rated residual operating current not exceeding 30 mA.

13 Following major repair to an electrical appliance, it is recommended that the appliance is subjected to

- ○ a   a user inspection
- ○ b   a consumer inspection
- ◉ c   production testing
- ○ d   type testing to an appropriate standard.

Answer c

Production testing is applied to new equipment and may also be appropriate to equipment following refurbishment or repair (see Section 6, The electrical tests). However, with reference to Appendix IV (Production testing), it should be remembered that production tests may be relatively arduous and as such should only be applied to items in as-new condition.

Option a, a user inspection, is a visual inspection carried out by the user at the intervals recommended in Table 7.1 in Section 7. It is covered in Section 13 (The user check). Option d, type testing, is carried out by manufacturers and testing houses to see if items comply with a standard and may be destructive in nature. As such, type testing is not appropriate for items in service. Option b, a consumer inspection, is not a topic covered by the Code of Practice.

From Section 6.3 Manufacturers carry out production testing to ensure that appliances are in accordance with the appropriate standard.

New appliances uncontaminated by dust or lubricants are subjected to production tests. Equipment in new or as-new condition may also be subjected to production testing following refurbishment or repair.

14 Class III equipment relies on electric shock protection by means of a SELV supply. A SELV supply shall <u>not</u> exceed

○ a 12 V a.c.
○ b 24 V a.c.
◉ c 50 V a.c.
○ d 120 V a.c.

Answer c

SELV sources should not exceed 50 V a.c. (see Section 11.3, Class III).

Both options a and b are less than 50 V a.c. and as such are also SELV. Option d is greater than 50 V a.c. and as such is classified as low voltage in *BS 7671: Requirements for Electrical Installations*.

From Section 11.3 SELV sources will not exceed 50 V a.c. and in many installations will be required to be below 24 or 12 V. SELV systems require specialist design and it is a requirement that there is no earth facility in the distribution of a SELV circuit or on the appliance or equipment.

Class III equipment is required to be supplied from a safety isolating transformer to BS EN 60742 or BS EN 61558-2-6.

15 A two-core extension lead <u>must</u> be fitted with

○ a an RCD
○ b a three-pin socket
◉ c a two-pin socket
○ d a BS EN 60309 plug.

Answer c

It is generally recommended that three-core leads (including a protective earthing conductor) be used. However, if a two-core cable is used, as in the question, it should never be connected to a standard 13 A three-pin plug socket (option b). This is to prevent the inadvertent use of such an unearthed extension lead to supply a piece of Class I equipment which relies on an effective earth conductor for protection against shock. See Section 5 (Types of electrical equipment) and Section 15.10 (Extension leads, multiway adaptors and RCD adaptors).

Options a and d are also not correct. The use of an RCD is advised only where extension leads exceed the lengths recommended and BS EN 60309 is the British Standard for plugs, socket-outlets and couplers for industrial purposes.

16 An item of equipment subjected to type testing will usually be

- ⦿ a   destroyed
- ○ b   sold as seen
- ○ c   used for spares
- ○ d   put into service.

Answer a

Appliances subjected to type testing are usually destroyed as the tests themselves are destructive. See Section 6.2 (Manufacturers' type testing).

Since damage may occur to equipment or components it would not be good practice to put the tested items into service (option d) or use them or their component parts for spares (option c). Option b, 'sold as seen', is a term more likely to be seen in a private advertisement selling an item and is not applicable here.

Section 6.2 Manufacturer's type testing   Test houses or the manufacturer carry out type testing to assess compliance with a standard (British or European). The tests are usually destructive, making that particular appliance unsuitable for sale or use.

17 A metal-cased item of equipment is supplied using a two-core cord set and has a classification mark as below. This item is

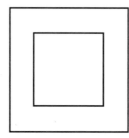

From the *IEE Code of Practice for In-service Inspection and Testing of Electrical Equipment*, Section 11.2 (Class II).

- ○ a   dangerous and should be removed from service
- ○ b   Class I
- ○ c   dangerous and should be fitted with a three-core cord
- ⦿ d   Class II.

Answer d

This symbol is the Class II construction mark. See Section 11.2 (Class II).

Class I items (option b) rely on an earth connection for reasons of safety and will usually have an earth symbol showing on the outer casing.

Outcome 1.3 In-service inspection and testing

18 A category of inspection and testing that may <u>not</u> require records to be kept is the

- ⦿ a user check
- ○ b production test
- ○ c formal visual inspection
- ○ d combined inspection and test.

Answer a

The Code of Practice refers to three categories of in-service inspection and testing: user checks, formal visual inspections and combined inspections and tests (options a, c and d). Only user checks do not always require records to be kept. See Section 7.2 (Categories of inspection and testing) and Section 13 (The user check).

Although production testing (option b) does not form part of in-service inspection and testing, evidence that the test was conducted is still required to prove that the equipment was safe following manufacture, usually in the form of a certificate of conformity.

Section 7.2 Three categories of in-service inspection and testing:
1 **User checks:** faults are to be reported and logged and faulty equipment should be removed from service. No record is required if no fault is found.
2 **Formal visual inspections:** inspections without tests, the results of which, satisfactory or unsatisfactory, are recorded.
3 **Combined inspections and tests:** the results of which are recorded.

19 The <u>most</u> important check to be carried out on any item of equipment in the workplace is the

- ○ a in-service test
- ○ b production test
- ⦿ c formal visual inspection
- ○ d combined inspection and test.

Answer c

A formal visual inspection is the most important check that can be carried out on a piece of equipment. See Section 6.4 (In-service inspection and testing) and Section 7.3 (Frequency of inspection and testing).

The visual inspection should be able to identify a number of faults that may not be picked up by an in-service test, a production test or the combined inspection and test (options a, b and d). Indeed, it may be

Notes

decided as a result of the findings of the visual inspection to not actually carry out any electrical tests as to do so may prove dangerous. A visual inspection is particularly suited to locating visually obvious defects.

From Section 7.3 The most important check that can be carried out on a piece of equipment is the visual inspection. The visual inspection can identify many defects, particularly in the case of portable appliances or hand-held tools where defects in the plug, the cable or the casing could occur.

20 Recorded tests may be omitted for Class II equipment

- ◉ a   located in an office
- ○ b   supplied using a three-core cord
- ○ c   located in a factory
- ○ d   supplied using a BS EN 60309 plug.

Answer a

If Class II equipment is located in a low-risk environment such as an office or hotel, recorded testing (but not inspection) may be omitted (see Table 7.1 in Section 7).

From the *IEE Code of Practice for In-service Inspection and Testing of Electrical Equipment*, Section 7, extract from Table 7.1.

Table 7.1   Initial frequency of inspection and testing of equipment

Equipment Use	Type of Equipment	User Checks	Class I		Class II	
			Formal visual inspection	Combined inspection and testing	Formal visual inspection	Combined inspection and testing
		Not recorded unless a fault is found	Recorded	Recorded	Recorded	Recorded
a	b	c	d	e	f	g
5 Hotels	S	none	24 months	48 months	24 months	none
	IT	none	24 months	48 months	24 months	none
	M	weekly	12 months	24 months	24 months	none
	P	weekly	12 months	24 months	24 months	none
	H	before use	6 months	12 months	6 months	none
6 Offices and shops	S	none	24 months	48 months	24 months	none
	IT	none	24 months	48 months	24 months	none
	M	weekly	12 months	24 months	24 months	none
	P	weekly	12 months	24 months	24 months	none
	H	before use	6 months	12 months	6 months	none

21 <u>Four</u> **factors that should influence the frequency of inspection and testing of appliances are the**

- ◉ a   environment, users, equipment construction and equipment type
- ◯ b   users, equipment construction, equipment type and cost involved
- ◯ c   equipment construction, equipment type, cost involved and environment
- ◯ d   equipment type, cost involved, environment and users.

Answer a

Section 7.3 (Frequency of inspection and testing) explains the four factors that should be considered.

Examiner's tip: With a question such as this, it is often easier to find the incorrect element within the answers. In this case, it is 'cost involved'.

From Section 7.3   Factors influencing the decision include the following:

1   **The environment:** equipment installed in a benign environment, such as an office, will suffer less damage than equipment in an arduous environment, such as a construction site.
2   **The users:** if the users of equipment report damage as and when it becomes evident, hazards will be avoided. Conversely, if equipment is likely to receive unreported abuse, more frequent inspection and testing is required.
3   **The equipment construction:** the safety of a Class I appliance is dependent upon a connection with the earth of the fixed electrical installation. If the flexible cable is damaged the connection with earth can be lost. The safety of Class II equipment is not dependent upon the electrical installation.
4   **The equipment type:** an appliance that is hand-held is more likely to be damaged than a fixed appliance. If such an appliance is also Class I the risk of danger is increased, as safety is dependent upon the continuity of the protective conductor from the plug to the appliance.

22 Table 7.1 in Section 7 of the *IEE Code of Practice for In-service Inspection and Testing of Electrical Equipment* provides guidance on initial frequencies of inspection and testing. The intervals between inspection and testing may be changed depending on

- ◯ a   the costs involved to carry out the tests
- ◯ b   the availability of the inspector
- ◉ c   a pattern of failure or damage being established
- ◯ d   a suitable time when the equipment is available.

Answer c

See Section 7.4 (Review of frequency of inspection and testing). The other factors of cost and availability are not relevant here.

Section 7.4 The intervals between checks, formal inspections and tests should be kept under review, particularly until patterns of failure or damage, if any, are determined.

Particularly close attention must be paid to initial checks, formal inspections and tests to see if there is a need to reduce the intervals or change the equipment or its use.

After the first few inspections and tests consideration should be given to increasing the intervals or reducing them.

23 Class I equipment is considered to be at a higher risk of damage or deterioration when

- ⊙ a located on a construction site
- ○ b used in a school
- ○ c located in a factory
- ○ d located in a hotel.

Answer a

Equipment installed in a benign environment (such as a school, factory or hotel – options b, c and d) will suffer less damage than equipment in an arduous environment. See Section 7.3 (Frequency of inspection and testing). Also note that in Table 7.1 in Section 7 a shorter interval between inspection and testing is suggested for construction site equipment.

24 Where the class of an item of equipment cannot be established, the suggested initial frequency of inspecting and testing the equipment shall be

- ○ a as Class II
- ⊙ b as Class I
- ○ c extended
- ○ d ignored.

Answer b

Class I items of equipment rely on the presence of an adequate protective conductor. As such, if any doubt exists as to the actual class of an item of equipment, confirmation of the adequacy of the protective conductor must be ensured. Therefore testing as Class I will be necessary. See Section 15.1 (Preliminary inspection).

Notes

From Section 15.1 If the Class of equipment is not known, it is to be tested as Class I.

25 A formal visual inspection of equipment will involve careful scrutiny of the appliance,

- ⦿ a   the cord set, the suitability of the appliance and the environment
- ○ b   the suitability of the appliance, the environment and the fixed electrical installation
- ○ c   the environment, the fixed electrical installation and the cord set
- ○ d   the fixed electrical installation, the cord set and the suitability of the appliance.

Answer a

In Section 14 (The formal visual inspection), in-depth reference is made to the importance of assessing the condition of the equipment including the flexible connection, the particular hazards likely to be present in the work environment and the suitability of the item of equipment for the environment in which it is located.

The main incorrect element in options b, c and d, the fixed electrical installation, should be subjected to regular inspection and testing by suitably competent persons in accordance with the requirements of BS 7671: *Requirements for Electrical Installations* and subjected to maintenance as required.

26 Prior to a formal inspection of equipment, it should be established whether the equipment has any faults. The most appropriate person to ask is

- ○ a   a responsible person
- ○ b   a competent person
- ⦿ c   a regular user
- ○ d   the duty holder.

Answer c

A regular user of an item of equipment is the person most familiar with the equipment and as such may be the person best placed to make a judgement as to whether or not it is suitable for use. The regular user is also the person most likely to be aware of any reliability issues or instances of damage to the equipment and/or its flexible connection over time. See Section 13 (The user check).

27 Class I appliances must be tested for

- ⦿ a   earth continuity, insulation resistance (or touch current) and function
- ○ b   insulation resistance (or touch current), function and polarity
- ○ c   function, polarity and dielectric strength
- ○ d   polarity, dielectric strength and earth continuity.

Answer a

Class I appliances rely upon the presence of a protective conductor for reasons of safety, so a test for earth continuity will be necessary. See Section 2 (Definitions). It is also necessary to carry out tests for insulation resistance and functional checks (see Section 15.3, In-service tests).

28 Which test or inspection is <u>not</u> applicable to Class II appliance-couplers?

- ○ a   Insulation resistance
- ○ b   Corrosion
- ⦿ c   Earth continuity
- ○ d   Mechanical damage

Answer c

Class II items do not require a protective conductor for reasons of safety (see Section 2, Definitions). Also see Section 15.4 (The earth continuity test), which states that earth continuity testing should only be applied to Class I equipment/cords.

29 Which of the following tests may cause damage to vulnerable electronic equipment?

- ○ a   Earth continuity
- ○ b   Touch current measurement
- ○ c   Polarity
- ⦿ d   Insulation resistance

Answer d

The insulation resistance test involves applying a test voltage of 500 V d.c. through the item of equipment under test. This is typically more than twice the rated voltage and may cause damage to equipment or result in misleading results being obtained. To avoid damage occurring, it is important that the live conductors (that is, the phase and neutral conductors) are connected together when carrying out this test. This step will also prevent misleading test results being given. See Section 15.5 (The insulation resistance test).

Outcome 1.4 Procedures, documentation, etc

30 The <u>main</u> purpose of an equipment register is to

- ○ a   assist with stock control
- ○ b   identify corporate thefts
- ○ c   record maintenance details
- ◉ d   review the frequency of inspection and testing.

Answer d

The equipment register is a vital management tool for checking that all equipment within a premises is appropriately inspected and tested at regular intervals. See Section 8.3 (Documentation) and Form VI.1, Equipment register, in Appendix VI.

31 Which of the following should appear on the label applied to equipment after testing?

- ◉ a   Re-testing due date
- ○ b   Previous test result data
- ○ c   Inspector's company name and logo
- ○ d   Manufacturer's production test certificate

Answer a

A label should be applied to equipment after testing including such information as a unique identification code, an indication of its current safety status, and the date on which re-testing is due or the last test date. See Section 8.4 (Labelling).

From the *IEE Code of Practice for In-service Inspection and Testing of Electrical Equipment*, Appendix VI, Form VI.3.

32 The model form that does <u>not</u> require the equipment serial number to be recorded is the

- ○ a   equipment register (Form VI.1)
- ○ b   inspection and test record (Form VI.2)
- ○ c   repair register (Form VI.4)
- ◉ d   faulty equipment register (Form VI.5).

Notes

Answer d

There is a space to fill in the serial number of a piece of equipment in Forms VI.1, VI.2 and VI.4, but no reference to serial number on Form V1.5.

33 A value of insulation resistance recorded as 1.34 MΩ is equivalent to

- ○ a   13,400 Ω
- ○ b   134,000 Ω
- ◉ c   1,340,000 Ω
- ○ d   13,400,000 Ω.

Answer c

1.34 MΩ is an abbreviation of 1.34 million ohms, or 1,340,000 Ω (option c). It is necessary to be familiar with the various ways of abbreviating and indicating results, as different instruments use different methods.

34 The nominal resistance for the protective conductor for a cord set having a cross-sectional area of 1 mm^2 and a length of 5 m is

- ◉ a   0.0975 Ω
- ○ b   0.975 Ω
- ○ c   9.750 Ω
- ○ d   97.50 Ω.

Answer a

In Section 15.4, Note (4), reference is made to Table VII.1 in Appendix VII (Nominal resistances of appliance supply cable protective conductors). An extract is shown below. For a 1 mm^2 cross-sectional area cable of length 5 m, resistance is stated as 97.5 mΩ, which is the same as 0.0975 Ω.

From the *IEE Code of Practice for In-service Inspection and Testing of Electrical Equipment*, Appendix VII, extract from Table VII.1.

Table VII.1   Nominal resistances of appliance supply cable protective conductors (Figures are for cables to BS 6500 or BS 6360)

Nominal conductor csa (mm^2)	Nominal conductor resistance at 20°C (mΩ/m)	Length (m)	Resistance at 20°C (mΩ)	Maximum current-carrying capacity (A)	Maximum diameter of individual wires in conductor (mm)	Approx. no. of wires in conductor
1.0	19.5	1	19.5	10	0.21	32
		1.5	29.3			
		2	39			
		2.5	48.8			
		3	58.5			
		4	78			
		5	**97.5**			

35 An instrument being used to carry out an earth continuity test should produce a test current of not less than 1.5 times the rating of the fuse and not greater than

○ a   100 mA
○ b   200 mA
◉ c   26 A
○ d   100 A.

Answer c
Section 15.4 (The earth continuity test) describes two test methods that may be adopted. One of these methods, the 'hard' test, states a maximum test current value of 26 A, so option c is correct.

From Section 15.4 Hard test. A continuity measurement should be made with a test current not less than 1.5 times the rating of the fuse up to a maximum of the order of 26 A for a period of between 5 and 20 seconds.

36 If a low resistance ohmmeter is used for earth continuity testing, a suitable test current would be

○ a   0.01 mA
○ b   10 mA
◉ c   100 mA
○ d   500 mA.

Answer c
Two test methods are described in Section 15.4 (The earth continuity test). One method, the 'soft' test requires a continuity measurement to be carried out with a short-circuit test current within the range 20 mA to 200 mA. Of the four answers offered, only answer c (100 mA) falls within this range.

37 A test facility, found on certain appliance test instruments, which is a useful means of determining whether or not a dual element heater is fully functioning, is the

○ a   dielectric strength test
◉ b   load test
○ c   earth continuity test
○ d   insulation resistance test.

Answer b
Load testing is a useful means of identifying whether one or more heating elements within an item being tested are open circuit. See Section 15.7 (Functional checks).

Notes

From Section 15.7 The use of more sophisticated instruments may permit load testing, which is an effective way of determining whether there are certain faults in appliances. It is particularly useful for heating appliances and will identify whether one or more elements are open circuit.

38 Any equipment that is faulty due to being unsuitable for the use intended should be

- ○ a   repaired
- ○ b   tested more frequently
- ○ c   replaced with an identical item
- ◉ d   replaced with a suitable item.

Answer d

Items that are not suitable for use (or for use within a particular location) should be removed and replaced by suitable equipment. See Section 8.5 (Damaged or faulty equipment).

From Section 8.5 If equipment is found to be damaged or faulty on inspection or test, it must be removed from service and then an assessment should be made by a responsible person as to the suitability of the equipment for the use in that particular location. ... If it is unsuitable it should be replaced by more suitable equipment.

39 An item of equipment is found to be unsafe and is reported to the responsible person by the inspector. The correct immediate action is to

- ○ a   have the item inspected more frequently
- ◉ b   remove the equipment from use
- ○ c   fix a failure label to the equipment
- ○ d   ask the user to take care when using the equipment.

Answer b

Section 8.5 (Damaged or faulty equipment) states that the responsible person should remove equipment from use immediately and then assess its suitability for future use.

Outcome 1.5 Training

40 It is recommended that users of appliances receive suitable training in order to carry out

- ○ a   formal inspections
- ○ b   combined inspections and tests
- ◉ c   user checks and reports of any faults found
- ○ d   suitable repairs to faulty equipment.

Answer c

Users of equipment are well placed to comment on the condition of the equipment that they operate on a regular basis and to carry out user checks. However, it may be necessary to prepare the users of equipment with appropriate training to enable them to carry out these user checks properly. See Section 13 (The user check), which states the actions to be taken when items of equipment are identified as being faulty.

41 Managers of testing organisations should receive training in order to understand their legal responsibilities as laid down in the

- ◉ a   Electricity at Work Regulations 1989
- ○ b   Health and Safety at Work etc. Act 1974
- ○ c   Electricity Safety, Quality and Continuity Regulations 2002
- ○ d   *BS 7671: Requirements for Electrical Installations.*

Answer a

Managers and administrators of premises and of inspection and testing organisations must be trained so that they are aware of their legal responsibilities as laid down in the Electricity at Work Regulations 1989. They should also understand Sections 1 to 8 of the *Code of Practice for In-service Inspection and Testing of Electrical Equipment*. See Section 9 (Training).

Outcome 1.6 Test instruments

42 The <u>most</u> suitable instrument for conducting an earth continuity test on an item of equipment that is connected directly to a switched fused connection unit is

○ a   a low resistance ohmmeter/insulation resistance tester set to 500 V

◉ b   a low resistance ohmmeter/insulation resistance tester set to low ohms

○ c   a portable appliance test instrument having a 13 A plug and socket facility

○ d   an earth loop impedance tester set at 20 Ω.

Answer b

In the case of items of equipment connected via a switched, fused connection unit no plug is fitted and so connection to the spring clips of a low resistance ohmmeter may be more convenient than using a portable appliance tester, which in most cases are best suited for testing items of equipment with some form of plug fitted. See Section 10.3 (Low resistance ohmmeters (for earth continuity testing)).

From Section 10.3   Earth continuity testing may in certain circumstances ... be carried out by a low resistance ohmmeter.

43 It is recommended that an insulation resistance tester applies a test voltage of

○ a   500 V a.c.

◉ b   500 V d.c.

○ c   3750 V a.c.

○ d   3750 V d.c.

Answer b

The applied test voltage should be approximately 500 V d.c. (see Section 15.5, The insulation resistance test). This is consistent with the requirement given in *BS 7671: Requirements for Electrical Installations* for insulation resistance testing of circuits in fixed electrical installations of nominal voltage up to and including 500 V.

44 A suitable alternative test for insulation resistance, where a high voltage may damage equipment, is

Notes

○ a   an earth continuity test
○ b   a load test
○ c   a functional test
◉ d   a measured touch current test.

Answer d

A protective conductor/touch current measurement can be used as an alternative to insulation resistance testing where such a test either cannot be carried out (because it may cause damage to equipment) or may result in a misleading result being obtained. See Section 15.6 (Protective conductor/touch current measurement).

45 In order to avoid having to re-test a large number of appliances, it is recommended that test equipment is

○ a   replaced yearly
◉ b   regularly checked for accuracy
○ c   the most modern model
○ d   able to perform touch current testing.

Answer b

All instruments should be subjected to regular checks for accuracy. This avoids certification being issued containing inaccurate test results and the subsequent need to re-test appliances in order to ascertain at what point the instrument(s) became inaccurate. See Section 10.6 (Instrument accuracy).

Answer key

Sample test

Question	Answer	Question	Answer
1	c	31	a
2	b	32	d
3	d	33	c
4	b	34	a
5	a	35	c
6	b	36	c
7	c	37	b
8	d	38	d
9	c	39	b
10	c	40	c
11	b	41	a
12	c	42	b
13	c	43	b
14	c	44	d
15	c	45	b
16	a		
17	d		
18	a		
19	c		
20	a		
21	a		
22	c		
23	a		
24	b		
25	a		
26	c		
27	a		
28	c		
29	d		
30	d		

Exam practice 2: Inspection and testing (2377–200)

Sample test 1   64

Questions and answers   71

Answer key   86

Exam practice 2

Sample test 1

The sample test below has 30 questions, which is the same number as the online Inspection and Testing of Electrical Equipment exam (2377–200), and its structure follows that of the online exam. The test appears firstly without answers, so you can use it as a mock exam. It is then repeated with answers and explanations. Finally, there is an answer key for easy reference.

Answer the questions by filling in the circle next to your chosen option.

Outcome 2.1   Equipment construction

1   **An electric heater has a mass of 16 kg and has castors to facilitate movement. The equipment type is**

○ a   fixed
○ b   transferable
○ c   movable
○ d   hand-held.

2   **A metal-cased electric iron has basic, as well as supplementary, insulation around live parts and is supplied by a two-core cord. The appliance construction is classified as Class**

○ a   I
○ b   II
○ c   III
○ d   0.

3   **An item of equipment that has the classification mark shown below must have**

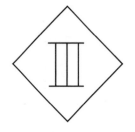

○ a   an earth connection
○ b   a three-core cord set
○ c   a 12 V supply
○ d   no earth connection.

4 Contact with an exposed-conductive-part made live by a fault will result in an electric shock due to a failure of

- ○ a   basic protection
- ○ b   mechanical protection
- ○ c   fault protection
- ○ d   ingress protection.

5 A Class II appliance has small metallic parts, which can be touched during operation. It is permissible for these parts to be isolated from any live part, by using only

- ○ a   basic insulation
- ○ b   supplementary insulation
- ○ c   reinforced insulation
- ○ d   earthed metallic parts.

6 If the resistance of the protective conductor core for a 2 m length of 0.75 mm² flex is 0.052 Ω, the resistance of a 1.5 mm² earth core of the same length would be

- ○ a   0.026 Ω
- ○ b   0.052 Ω
- ○ c   0.104 Ω
- ○ d   0.152 Ω.

Outcome 2.2 Inspection

7 For portable equipment used by school children, it is recommended that initial checks are made on a weekly basis. These checks should be carried out by

- ○ a   the teacher or supervisor
- ○ b   a competent inspector
- ○ c   the user
- ○ d   the local authority.

8 The type of equipment that requires <u>more</u> frequent initial formal visual inspections in a hotel is

- ○ a   information technology equipment
- ○ b   stationary equipment
- ○ c   portable equipment
- ○ d   hand-held equipment.

9   An insulation resistance test on a Class II hand-held item of equipment produced a measurement of 20 MΩ. A test six months later, on the same item of equipment, produced a measurement of 5 MΩ. This would indicate that the equipment is

○ a   more efficient
○ b   earthed more effectively
○ c   deteriorating
○ d   of no concern at all.

10 Which one of the following would <u>not</u> be considered part of a user check? Checking

○ a   the flex or appliance cord condition
○ b   the socket-outlet for signs of overheating
○ c   the socket-outlet for polarity
○ d   that the equipment is suitable for the environment.

11 A measurement of earth continuity for Class I equipment should not exceed

○ a $0.1\,\Omega + R$
○ b $0.5\,\Omega + R$
○ c $1.0\,\Omega + R$
○ d $2.0\,\Omega + R.$

12 A label attached to equipment that has passed all the required inspection and testing should <u>not</u> contain the words

○ a   pass
○ b   do not use
○ c   date of check
○ d   next test before.

Outcome 2.3   Combined inspection and testing

13 Before a formal visual inspection is carried out on business equipment, care must be taken that the equipment is disconnected from the normal electrical supply and

○ a   any wireless network
○ b   any uninterruptible power supply
○ c   associated keyboards
○ d   wireless modems.

14 A means of isolating portable appliances

○ a   is not necessary
○ b   should always be by a plug and socket-outlet
○ c   should be readily accessible by the user
○ d   should only be accessible by skilled persons.

15 The cord anchorage of a plug must always secure the

○ a   phase and neutral cores
○ b   phase and earth cores
○ c   phase, neutral and earth cores
○ d   cable sheath.

16 A visual inspection of the face plate of a socket-outlet may reveal

○ a   signs of overheating
○ b   incorrect polarity
○ c   poor earth continuity
○ d   incorrect fusing.

17 Any non-safety earthed metal parts should be tested for earth continuity by using a

○ a   25 A test current for 20 seconds
○ b   low current continuity tester
○ c   test current 1.5 times the fuse rating
○ d   load/leakage test as an alternative.

18 In order to protect the equipment under test whilst conducting an insulation resistance test, it is important to connect the

○ a   phase and earth conductors
○ b   neutral and earth conductors
○ c   phase and neutral conductors
○ d   phase, neutral and earth conductors.

Outcome 2.4 Use of instruments, etc

19 A suitable test to determine whether a heating appliance with multiple elements operates correctly is

○ a   an insulation resistance test
○ b   an earth continuity test
○ c   a load test
○ d   a touch current measurement.

Notes

20 Dielectric strength testing is <u>not</u> recommended as part of in-service inspection and testing as it may damage

- ○ a   insulation
- ○ b   conductors
- ○ c   the test instrument
- ○ d   the fixed installation.

21 Earth continuity may be tested using a portable appliance test instrument (PAT) or

- ○ a   an insulation resistance tester
- ○ b   an earth electrode resistance tester
- ○ c   an earth loop impedance tester
- ○ d   a low resistance ohmmeter.

22 Deterioration to an earth connection may be occurring if previous earth continuity test results recorded a reading that was, when compared to current results,

- ○ a   the same
- ○ b   slightly higher
- ○ c   considerably lower
- ○ d   1.5 times higher.

23 Measured touch current readings taken during a test on a Class II appliance must not exceed

- ○ a   0.25 mA
- ○ b   3.5 mA
- ○ c   0.25 A
- ○ d   3.5 A.

24 When reviewing intervals between checks beyond the initial inspection/testing, the <u>most</u> important information would be

- ○ a   the previous inspection and test results
- ○ b   the equipment registers
- ○ c   Table 7.1 in Section 7 of the *IEE Code of Practice for In-service Inspection and Testing of Electrical Equipment*
- ○ d   the equipment serial number.

Outcome 2.5 Equipment

25 An extension lead fitted with a three-pin socket-outlet should be tested as a Class I appliance, with the addition of

- ○ a   a polarity check
- ○ b   a dielectric strength test
- ○ c   an earth leakage test
- ○ d   a load test.

26 Where a 0.5 mm² appliance flex is protected by a fuse in a BS 1363 plug, the fuse rating should not exceed

- ○ a   3 A
- ○ b   5 A
- ○ c   10 A
- ○ d   13 A.

27 Where a single item of information technology equipment has a stated touch current of 4.7 mA, it shall be supplied from the fixed electrical installation by a

- ○ a   BS EN 60309-2 plug
- ○ b   BS 1363 plug
- ○ c   BS 3535 transformer
- ○ d   BS EN 60950 supply.

28 A 230 V appliance has an input rating of 550 W. For standardisation, the correct fuse size for the three-pin plug would be

- ○ a   1 A
- ○ b   3 A
- ○ c   5 A
- ○ d   13 A.

29 Where an item of equipment has undergone extensive repairs or refurbishment, the equipment may need to be subjected to

- ○ a   type testing
- ○ b   user checks
- ○ c   production testing
- ○ d   acoustic testing.

30 When conducting a functional test on a microwave oven, opening the door during use will

- ○ a   interrupt the oven power
- ○ b   operate the circuit-breaker
- ○ c   cause the audible alarm to sound
- ○ d   set the cooking timer to zero.

Questions and answers

The questions in sample test 1 are repeated below with worked-through answers and extracts from the *IEE Code of Practice for In-service Inspection and Testing of Electrical Equipment* where appropriate. Where references to sections are made and extracts given, these may be found in the Code of Practice publication.

Outcome 2.1 Equipment construction

1 **An electric heater has a mass of 16 kg and has castors to facilitate movement. The equipment type is**

○ a   fixed
○ b   transferable
◉ c   movable
○ d   hand-held.

Answer c
An appliance of less than 18 kg in mass and not fixed, or equipment fitted with wheels or castors, is categorised as movable equipment. See Section 5 (Types of electrical equipment).

From Section 5.2
An item of movable equipment is equipment which is either:
• 18 kg or less in mass and not fixed, eg an electric compressor, or
• equipment with wheels, castors or other means to facilitate movement by the operator as required to perform its intended use, eg air-conditioning unit.

2 **A metal-cased electric iron has basic, as well as supplementary, insulation around live parts and is supplied by a two-core cord. The appliance construction is classified as Class**

○ a   I
◉ b   II
○ c   III
○ d   0.

Answer b
See Section 2 (Definitions) for a definition of Class II equipment, as well as definitions of Class I, III and 0.

Notes

From the *IEE Code of Practice for In-service Inspection and Testing of Electrical Equipment*, Section 11.3

From Section 2   Class II equipment. Equipment in which protection against electric shock does not rely on basic insulation only, but in which additional safety precautions such as supplementary insulation are provided, there being no provision for the connection of exposed metalwork of the equipment to a protective conductor and no reliance upon precautions to be taken in the fixed wiring of the installation.

3   **An item of equipment that has the classification mark shown below must have**

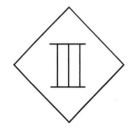

- ○ a   an earth connection
- ○ b   a three-core cord set
- ○ c   a 12 V supply
- ◉ d   no earth connection.

Answer d
The symbol shown is the Class III construction mark. See Section 11.3 (Class III). To protect against electric shock, Class III equipment relies on supply from a SELV source that is electrically separated from earth.

From Section 2   Class III equipment. Equipment in which protection against electric shock relies on a supply at SELV and in which voltages higher than those of SELV are not generated.

4   **Contact with an exposed-conductive-part made live by a fault will result in an electric shock due to a failure of**

- ○ a   basic protection
- ○ b   mechanical protection
- ◉ c   fault protection
- ○ d   ingress protection.

Answer c
Should a fault occur in a Class I appliance, fault current would normally flow to earth through the circuit protective conductor. If there is a failure in fault protection, eg the circuit protective conductor has become disconnected from the appliance, the exposed-conductive-parts would

then become live as there would be no direct means of earthing. This would then result in an electric shock should the appliance be touched.

Option a, basic protection, is incorrect as *BS 7671: Requirements for Electrical Installations* defines it as 'protection against electric shock under fault-free conditions'. Option b, mechanical protection, is incorrect as it refers to protecting electrical installations from impact damage and option d, ingress protection, is incorrect as it refers to the IP rating of electrical equipment.

5 **A Class II appliance has small metallic parts, which can be touched during operation. It is permissible for these parts to be isolated from any live part, by using only**

○ a   basic insulation
○ b   supplementary insulation
◉ c   reinforced insulation
○ d   earthed metallic parts.

Answer c
It is acceptable for small metallic parts to be isolated from live parts by insulation at least equivalent to reinforced insulation. See Section 2 (Definitions, Class II equipment). Such parts are unlikely to become live under fault conditions as a result of their size and the standard of insulation between them and live parts.

6 **If the resistance of the protective conductor core for a 2 m length of 0.75 mm² flex is 0.052 Ω, the resistance of a 1.5 mm² earth core of the same length would be**

◉ a   0.026 Ω
○ b   0.052 Ω
○ c   0.104 Ω
○ d   0.152 Ω.

Answer a
There are two ways of answering this question. One way is to refer to Table VII.1 in Appendix VII, which states that a 2 m length of 1.5 mm² protective conductor has a resistance of 26.6 mΩ. This is approximately the same as 0.026 Ω, which is the answer. The correct answer can also be obtained by halving the resistance of 0.052 Ω, stated as the resistance of the 0.75 mm² protective conductor in the question. As the cross-sectional area of the conductor has doubled, its resistance will be halved.

Notes

From the *IEE Code of Practice for In-service Inspection and Testing of Electrical Equipment*, Appendix VII, extract from Table VII.1.

Table VII.1 Nominal resistances of appliance supply cable protective conductors (Figures are for cables to BS 6500 or BS 6360)

Nominal conductor csa (mm²)	Nominal conductor resistance at 20°C (mΩ/m)	Length (m)	Resistance at 20°C (mΩ)	Maximum current-carrying capacity (A)	Maximum diameter of individual wires in conductor (mm)	Approx. no. of wires in conductor
1.5	13.3	1	13.3	15	0.26	30
		1.5	20			
		2	**26.6**			
		2.5	33.3			
		3	39.9			
		4	53.2			
		5	66.5			

Outcome 2.2 Inspection

7 **For portable equipment used by school children, it is recommended that initial checks are made on a weekly basis. These checks should be carried out by**

- ⦿ a   the teacher or supervisor
- ○ b   a competent inspector
- ○ c   the user
- ○ d   the local authority.

Answer a

In the case of equipment in schools any user checks should be carried out by a supervisor, teacher or member of staff. Table 7.1 in Section 7 states the suggested frequencies for user checks, inspection and testing.

8 **The type of equipment that requires <u>more</u> frequent initial formal visual inspections in a hotel is**

- ○ a   information technology equipment
- ○ b   stationary equipment
- ○ c   portable equipment
- ⦿ d   hand-held equipment.

Notes

Answer d

Referring to Table 7.1 in Section 7, it can be seen that hand-held equipment (H) requires formal visual inspections most frequently.

Table 7.1 Initial frequency of inspection and testing of equipment

	Equipment Use	Type of Equipment	User Checks	Class I		Class II	
			Not recorded unless a fault is found	Formal visual inspection	Combined inspection and testing	Formal visual inspection	Combined inspection and testing
				Recorded	Recorded	Recorded	Recorded
	a	b	c	d	e	f	g
5	Hotels	S	none	24 months	48 months	24 months	None
		IT	none	24 months	48 months	24 months	None
		M	weekly	12 months	24 months	24 months	None
		P	weekly	12 months	24 months	24 months	None
		H	before use	**6 months**	12 months	**6 months**	None

From the *IEE Code of Practice for In-service Inspection and Testing of Electrical Equipment*, Section 7, extract from Table 7.1.

9 **An insulation resistance test on a Class II hand-held item of equipment produced a measurement of 20 MΩ. A test six months later, on the same item of equipment, produced a measurement of 5 MΩ. This would indicate that the equipment is**

- ○ a more efficient
- ○ b earthed more effectively
- ◉ c deteriorating
- ○ d of no concern at all.

Answer c

A reduction in the value of insulation resistance achieved would be an indication of deterioration of the insulation. An insulation resistance test on a Class II item is carried out to confirm that live (ie phase and neutral) conductors are sufficiently electrically separate from each other. With reference to Table 15.2 (Insulation resistance readings) in Section 15, if the insulation resistance between the live conductors in a Class II item of equipment falls below 2 MΩ, it should be withdrawn from service.

10 Which one of the following would <u>not</u> be considered part of a user check? Checking

○ a   the flex or appliance cord condition
○ b   the socket-outlet for signs of overheating
◉ c   the socket-outlet for polarity
○ d   that the equipment is suitable for the environment.

Answer c

Options a, b and d, checking flex or cord condition, signs of overheating on the socket-outlet and suitability for the environment, are all checks that may be made by the user. Therefore, option c is the answer. See Section 13 (The user check).

Examiner's tip: Ensure you read each question carefully as this type of negative question, eg 'which one is **not**…' is sometimes confusing and can lead to the incorrect answer being given.

11 A measurement of earth continuity for Class I equipment should not exceed

◉ a $0.1\,\Omega + R$
○ b $0.5\,\Omega + R$
○ c $1.0\,\Omega + R$
○ d $2.0\,\Omega + R$.

Answer a

The earth continuity for Class I equipment should not exceed $(0.1 + R)\,\Omega$, so only option a is correct. See Section 15.4 (The earth continuity test).

12 A label attached to equipment that has passed all the required inspection and testing should <u>not</u> contain the words

○ a   pass
◉ b   do not use
○ c   date of check
○ d   next test before.

Answer b

The label should contain the following information: indication that it has passed the testing, a unique identification code and the date on which the re-testing is due or the last test date. See Section 8.4 (Labelling).

Option b, 'do not use', would only be included if the item being tested was considered unsuitable for continued use as a result of the inspection/testing carried out upon it.

Outcome 2.3 Combined inspection and testing

13 Before a formal visual inspection is carried out on business equipment, care must be taken that the equipment is disconnected from the normal electrical supply and

- ○ a   any wireless network
- ◉ b   any uninterruptible power supply
- ○ c   associated keyboards
- ○ d   wireless modems.

Answer b

When equipment is isolated, it is necessary for it to be disconnected from all sources of supply. The person carrying out the formal visual inspection should confirm that equipment is isolated from the normal supply and any uninterruptible power supplies (or indeed other forms of standby supply) prior to commencing the inspection. The only means of supply listed as an option in the question is an uninterruptible power supply, therefore this is the correct answer. See Section 14 (Formal visual inspection).

14 A means of isolating portable appliances

- ○ a   is not necessary
- ○ b   should always be by a plug and socket-outlet
- ◉ c   should be readily accessible by the user
- ○ d   should only be accessible by skilled persons.

Answer c

The means of isolation from the supply must be readily accessible to the user of the equipment (see Section 14, The formal visual inspection). If equipment is supplied via a plug and socket, it should be possible to access the socket-outlet in order to remove the plug. Some equipment may be supplied via a disconnector (also known as an isolator) or a switched, fused connector.

In order to isolate the equipment, the disconnector or switch on the connection unit should be placed in the 'off' position. If the means of isolation is not under the immediate control of the person carrying out the inspection, it should be secured in the 'off' position.

Notes

From the *IEE Code of Practice for In-service Inspection and Testing of Electrical Equipment*, Appendix VIII.

15 The cord anchorage of a plug must always secure the

○ a   phase and neutral cores
○ b   phase and earth cores
○ c   phase, neutral and earth cores
◉ d   cable sheath.

Answer d

The purpose of the cord anchorage, whether fitted in the plug or where the flexible cable enters an item of equipment, is to secure the cable. This is so that, in the event of mechanical strain being put on the cable, it does not have a detrimental effect on the conductors. This is achieved by clamping the outer sheath leaving the conductors themselves free of strain. See Section 13 (The user check).

The cable conductors are not designed to take any strain or mechanical loading. They are sized to carry the current required by the equipment in normal use (the design current). If subjected to undue strain, the conductors may become stretched, or individual strands within the conductor may be broken, resulting in a reduction of current-carrying capacity which could cause the overheating of the conductor.

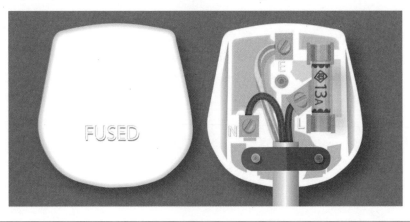

16 A visual inspection of the face plate of a socket-outlet may reveal

◉ a   signs of overheating
○ b   incorrect polarity
○ c   poor earth continuity
○ d   incorrect fusing.

Answer a
It is possible to detect many signs of damage or indicators of general condition and serviceability by carrying out a visual inspection. In the case of socket-outlets, it is mentioned that the inspector should look for cracks, other damage and signs of overheating. See Section 13 (The user check).

17 Any non-safety earthed metal parts should be tested for earth continuity by using a

- a   25 A test current for 20 seconds
- ⦿ b   low current continuity tester
- c   test current 1.5 times the fuse rating
- d   load/leakage test as an alternative.

Answer b
It is recommended that connections are checked using a low current continuity tester to perform a continuity test method carried out with a test current of between 20 mA and 200 mA. These non-safety earthed metal parts must not be subjected to a continuity measurement with a 25 A test current as doing so may result in damage occurring to the equipment. See Section 15.4 (The earth continuity test).

From Section 15.4, Note 1 Some equipment may have accessible metal parts that are earthed only for functional or screening purposes, with protection against electric shock being provided by double or reinforced insulation... Connections may be checked using a low current continuity test instrument.

18 In order to protect the equipment under test whilst conducting an insulation resistance test, it is important to connect the

- a   phase and earth conductors
- b   neutral and earth conductors
- ⦿ c   phase and neutral conductors
- d   phase, neutral and earth conductors.

Answer c
When an insulation resistance test is carried out on an item of equipment, typically a test voltage of 500 V d.c. is applied. This is generally considerably higher than the normal rated operating voltage of the equipment or the components therein, some of which may be voltage sensitive and therefore vulnerable to being damaged when the insulation resistance test is performed. To remove the possibility of damage occurring, the insulation resistance may be measured between live

conductors (that is, with the phase and neutral connected together) and earth. By so doing, there is no potential difference (voltage) applied between phase and neutral and thereby across components. This test procedure is described in Section 15.5 (The insulation resistance test).

Outcome 2.4   Use of instruments, etc

19 A suitable test to determine whether a heating appliance with multiple elements operates correctly is

- ○ a   an insulation resistance test
- ○ b   an earth continuity test
- ◉ c   a load test
- ○ d   a touch current measurement.

Answer c
Load testing is a useful means of identifying whether one or more heating elements within an item being tested are open circuit. See Section 15.7 (Functional checks).

From Section 15.7   The use of more sophisticated instruments may permit load testing, which is an effective way of determining whether there are certain faults in appliances. It is particularly useful for heating appliances and will identify whether one or more elements are open circuit.

20 Dielectric strength testing is <u>not</u> recommended as part of in-service inspection and testing as it may damage

- ◉ a   insulation
- ○ b   conductors
- ○ c   the test instrument
- ○ d   the fixed installation.

Answer a
Dielectric strength testing may result in damage occurring to insulation or electronic devices/circuits (see Section 15.3, In-service tests). Dielectric testing is also not to be carried out as part of in-service testing, but is normally carried out by a manufacturer on a new appliance.

From Section 15.3 … equipment should not normally be subjected to dielectric strength testing (also known as flash testing or hi-pot testing) because this may damage insulation and may also indirectly damage low voltage electronic circuits unless appropriate precautions are taken.

Notes

21 Earth continuity may be tested using a portable appliance test instrument (PAT) or

○ a   an insulation resistance tester
○ b   an earth electrode resistance tester
○ c   an earth loop impedance tester
◉ d   a low resistance ohmmeter.

Answer d
Earth continuity can be tested using a low resistance ohmmeter (see Section 10.3, Low resistance ohmmeters (for earth continuity testing)). In the case of items of equipment connected via a switched fused connection unit, no plug is fitted and so connection to the spring clips of a low resistance ohmmeter may be more convenient than using a portable appliance tester, which in most cases is best suited for testing items of equipment with some form of plug fitted.

From Section 10.3 Earth continuity testing may in certain circumstances be carried out by a low resistance ohmmeter.

22 Deterioration to an earth connection may be occurring if previous earth continuity test results recorded a reading that was, when compared to current results,

○ a   the same
○ b   slightly higher
◉ c   considerably lower
○ d   1.5 times higher.

Answer c
Generally, the continuity of an earth connection should be as low as possible and a measure of its reliability over time would be for it to remain substantially unchanged. If, on examination of test results, the resistance of an earth connection is seen to be increasing, this would indicate a deterioration of the connection, or a loose connection. As such, the earlier reading should be lower than the current one.

23 Measured touch current readings taken during a test on a Class II appliance must not exceed

◉ a   0.25 mA
○ b   3.5 mA
○ c   0.25 A
○ d   3.5 A.

Answer a

The measured touch current of Class II equipment should not exceed 0.25 mA. See Table 15.3 in Section 15.6 (Protective conductor/touch current measurement).

Table 15.3 Measured protective conductor/touch current

Appliance Class	Maximum protective conductor/touch current
Class II equipment	0.25 mA

24 When reviewing intervals between checks beyond the initial inspection/testing, the <u>most</u> important information would be

- ⊙ a   the previous inspection and test results
- ○ b   the equipment registers
- ○ c   Table 7.1 in Section 7 of the *IEE Code of Practice for In-service Inspection and Testing of Electrical Equipment*
- ○ d   the equipment serial number.

Answer a

Previous test results should be examined to determine the effectiveness of the current inspection/testing regime and indeed to decide on the suitability (or otherwise) of equipment for its current use/location. The importance of previous test results in the review process is mentioned in Section 7.4 (Review of frequency of inspection and testing), Section 8.4 (Documentation) and Section 15 (Combined inspection and testing).

Outcome 2.5 Equipment

25 An extension lead fitted with a three-pin socket-outlet should be tested as a Class I appliance, with the addition of

- ⊙ a   a polarity check
- ○ b   a dielectric strength test
- ○ c   an earth leakage test
- ○ d   a load test.

Answer a

Extension leads fitted with a standard three-pin socket-outlet should be tested as Class I appliances with the addition of a polarity check. See Section 15.10.1 (Extension leads). Options b, c and d are not appropriate.

26 Where a 0.5 mm² appliance flex is protected by a fuse in a BS 1363 plug, the fuse rating should not exceed

◉ a   3 A
○ b   5 A
○ c   10 A
○ d   13 A.

Answer a

Flexes down to minimum cross-sectional area (csa) of 0.5 mm² may be protected by a 3 A BS 1363 fuse. See Section 15.13 (Replacement of appliance flexes).

27 Where a single item of information technology equipment has a stated touch current of 4.7 mA, it shall be supplied from the fixed electrical installation by a

◉ a   BS EN 60309-2 plug
○ b   BS 1363 plug
○ c   BS 3535 transformer
○ d   BS EN 60950 supply.

Answer a

Equipment with a current exceeding 3.5 mA should be supplied with a BS EN 60309-2 plug. This is a requirement of *BS 7671: Requirements for Electrical Installations*. See Section 15.12 (High protective conductor currents) in the Code of Practice.

Option b, BS 1363, is the standard for 13 A plugs, socket-outlets, connection units and adaptors. Option c, BS 3535, is a partially replaced standard for isolating transformers and safety isolating transformers. Option d, BS EN 60950, is the specification for safety of information technology equipment including electrical business equipment.

From Section 15.12 ... equipment with a protective conductor current designed to exceed 3.5 mA should be permanently wired to the fixed installation, or be supplied by an industrial plug and socket to BS EN 60309-2.

28 A 230 V appliance has an input rating of 550 W. For standardisation, the correct fuse size for the three-pin plug would be

○ a   1 A
◉ b   3 A
○ c   5 A
○ d   13 A.

Answer b

For appliances up to 700 W a 3 A fuse is suitable. See Section 15.14 (Plug fuses). A 13 A fuse (option d) is appropriate for appliances over 700 W.

29 Where an item of equipment has undergone extensive repairs or refurbishment, the equipment may need to be subjected to

○ a   type testing
○ b   user checks
◉ c   production testing
○ d   acoustic testing.

Answer c

A repairer may wish to subject an appliance either to production tests or in-service tests. The repairer should base their decision of what degree of testing is required on the condition of the equipment and the nature of the repairs that have been carried out. See Section 6.5 (Testing after repair).

Option a, type testing, as carried out by manufacturers or test houses, usually results in damage sufficient to make equipment or appliances unsuitable for further use. Option b, user checks, are those checks carried out by regular users of equipment at a frequency in accordance with the recommendations given in Table 7.1 in Section 7 of the *IEE Code of Practice*. Option d, acoustic testing, is not covered in the Code of Practice.

30 When conducting a functional test on a microwave oven, opening the door during use will

◉ a   interrupt the oven power
○ b   operate the circuit-breaker
○ c   cause the audible alarm to sound
○ d   set the cooking timer to zero.

Answer a

Whenever an inspection/test of a microwave oven is carried out, a functional check confirming that the opening of the door results in interruption of the oven power should be performed. See Section 15.11 (Microwave ovens).

Answer key

Sample test 1

Question	Answer
1	c
2	b
3	d
4	c
5	c
6	a
7	a
8	d
9	c
10	c
11	a
12	b
13	b
14	c
15	d
16	a
17	b
18	c
19	c
20	a
21	d
22	c
23	a
24	a
25	a
26	a
27	a
28	b
29	c
30	a

Exam practice 3: Inspection and testing (2377–200)

Sample test 2	88
Questions and answers	95
Answer key	112

Exam practice 3

Notes

Sample test 2

The sample test below has 30 questions, which is the same number as the online Inspection and Testing of Electrical Equipment exam (2377–200), and its structure follows that of the online exam. The test appears firstly without answers, so you can use it as a mock exam. It is then repeated with answers and explanations. Finally, there is an answer key for easy reference.

Answer the questions by filling in the circle next to your chosen option.

Outcome 2.1   Equipment construction

1   **A microwave oven has a mass of 21 kg and is located on a kitchen worktop. The equipment type is**

○ a   portable
○ b   transportable
○ c   built in
○ d   stationary.

2   **A metal-cased electric iron has basic insulation around live parts and is supplied by a three-core cord. The appliance construction is Class**

○ a   I
○ b   II
○ c   III
○ d   01.

3   **Which <u>one</u> of the following classification marks indicates an item of equipment that does not rely on installation conditions for protection against electric shock?**

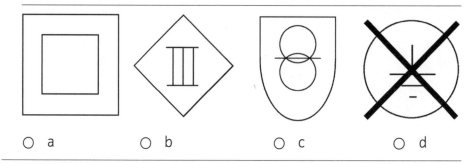

○ a   ○ b   ○ c   ○ d

4 **An electric shock through a failure of basic protection occurs when a person touches**

○ a   the metal casing of a Class I appliance made live by a fault
○ b   the earth terminal of an appliance in operation
○ c   the phase terminal of an appliance in operation, which has its
         outer covers removed
○ d   an exposed-conductive-part made live by a fault.

5 **In order for the appliance shown in the diagram below to be suitably protected in the event of a fault, the metal casing of the appliance must be**

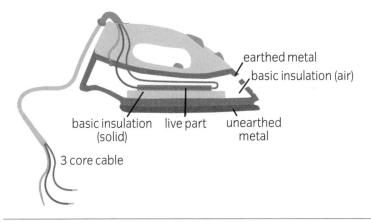

earthed metal
basic insulation (air)
basic insulation (solid)   live part   unearthed metal
3 core cable

○ a   connected to earth
○ b   disconnected from earth
○ c   reinforced
○ d   non-ferrous.

6 **If an appliance cord set length is doubled, the resistance value of the earth conductor will**

○ a   remain unchanged
○ b   decrease slightly
○ c   halve
○ d   double.

Notes

Outcome 2.2 Inspection

7 Portable equipment used in an office should be subjected to initial user checks

○ a   before use
○ b   daily
○ c   weekly
○ d   monthly.

8 Which classification and location of equipment from the following is considered to be of the lowest risk?

○ a   Class I equipment in an office
○ b   Class I equipment in a factory
○ c   Class II equipment in an office
○ d   Class II equipment on a construction site

9 Initial formal visual inspection on hand-held Class I equipment used by the public should be carried out

○ a   weekly
○ b   monthly
○ c   every 12 months
○ d   every 24 months.

10 An earth continuity test on an item of Class I equipment produced a result of 0.05 Ω. The same test on the same item of equipment was carried out six months later and produced a result of 0.46 Ω. This would indicate that the equipment has

○ a   good insulation resistance
○ b   a better earth connection
○ c   a higher touch current
○ d   a poorer earth connection.

11 The <u>most</u> appropriate class of equipment for use outdoors would be Class

○ a   I
○ b   II
○ c   0
○ d   01.

12 Any equipment showing a significant decrease in insulation resistance values, which remain within the allowed values, should be tested

- ○ a   more frequently
- ○ b   less frequently
- ○ c   by the manufacturer
- ○ d   for dielectric strength.

Outcome 2.3   Combined inspection and testing

13 An earth continuity test on a 22 m extension lead having a core area of 2.5 mm² would produce a result of approximately

- ○ a $0.017\,\Omega$
- ○ b $0.176\,\Omega$
- ○ c $0.821\,\Omega$
- ○ d $8.00\,\Omega$.

14 When conducting a formal visual inspection, care must be taken to ensure business equipment is isolated from normal electrical supply and any UPS connected. UPS is an abbreviation of

- ○ a   universal peripheral system
- ○ b   undervoltage protection system
- ○ c   universal provider supply
- ○ d   uninterruptible power supply.

15 A detachable appliance cord set for Class I equipment must

- ○ a   never have an earth core fitted
- ○ b   always be protected by a 13 A fuse
- ○ c   never have a non-rewireable type plug
- ○ d   always have a protective conductor connected.

16 Where a Class I item of equipment is <u>not</u> fitted with a plug, earth continuity testing may be easier using

- ○ a   a portable appliance test set with a 13 A socket
- ○ b   a continuity/insulation tester
- ○ c   an earth loop impedance tester
- ○ d   an RCD tester.

17 An insulation resistance test on a Class I appliance gives a test result of 0.02 MΩ. If the appliance is connected to a 230 V supply, the value of current that would leak to earth would be

- ○ a   1 mA
- ○ b   11.5 mA
- ○ c   100 mA
- ○ d   1 A.

18 If a touch current measurement is conducted as an alternative to insulation resistance testing, the test voltage used, for practical purposes, is

- ○ a   half the supply voltage
- ○ b   500 V a.c.
- ○ c   500 V d.c.
- ○ d   the supply voltage.

Outcome 2.4   Use of instruments, etc

19 A test for polarity is only required to be carried out on extension leads and

- ○ a   Class I hand-held appliances
- ○ b   non-detachable flexible cords
- ○ c   detachable flexible cords
- ○ d   Class II hand-held appliances.

20 Some electrical test equipment incorporates a test facility that is not recommended for in-service testing. This test facility is commonly known as

- ○ a   low pot testing
- ○ b   flash testing
- ○ c   flush testing
- ○ d   no pot testing.

21 Which of the following test equipment would be required to carry out in-service inspection and testing?

- ○ a   Earth loop impedance tester
- ○ b   Earth continuity tester
- ○ c   Prospective fault current tester
- ○ d   Clamp type ammeter

22 **An appliance is supplied with a 0.75 mm² flexible cord having a length of 2 m and an earth continuity test produces a reading of 0.172 Ω. The value of earth continuity for the appliance alone is**

○ a   0.052 Ω and acceptable
○ b   0.072 Ω and unacceptable
○ c   0.1 Ω and acceptable
○ d   0.12 Ω and unacceptable.

23 **Equipment labels should contain information such as**

○ a   date of check, safety status, appliance number and next test date
○ b   safety status, appliance number, next test date and test results
○ c   appliance number, next test date, test results and date of check
○ d   next test date, test results, date of check and safety status.

24 **Insulation deterioration may be occurring if previous insulation resistance test results recorded a reading that was, when compared to current results,**

○ a   much higher
○ b   the same
○ c   slightly lower
○ d   much lower.

Outcome 2.5   Equipment

25 **Any equipment found to be unsafe following an inspection and test shall be**

○ a   identified with a fail label and remain in use
○ b   identified with a pass label and withdrawn from use
○ c   identified with a fail label and withdrawn from use
○ d   identified with a pass label and remain in use.

26 **An extension lead has a cross-sectional area (csa) of 1.5 mm² and a length of 14 m. An earth continuity reading should give a result of approximately**

○ a   13.3 mΩ
○ b   19 mΩ
○ c   0.186 Ω
○ d   186.2 Ω.

Notes

27 The purpose of the fuse in a standard 13 A plug is to

- ○ a   protect the appliance
- ○ b   protect the fixed installation
- ○ c   conform with European standards
- ○ d   protect the flexible cord.

28 Equipment having a high protective conductor current exceeding 10 mA requires specific precautions to be observed. Further information on the arrangement of such equipment can be found in

- ○ a   BS 7176
- ○ b   GS38
- ○ c   BS 7671
- ○ d   PM28.

29 A warning label is required for any item of equipment having a protective conductor current in excess of

- ○ a   2.5 mA
- ○ b   3.5 mA
- ○ c   5 mA
- ○ d   6.5 mA.

30 A detachable, three-core cord set used to supply 230 V information technology equipment shall be tested as Class

- ○ a   I
- ○ b   II
- ○ c   III
- ○ d   0.

Questions and answers

The questions in sample test 2 are repeated below with worked-through answers and extracts from the *IEE Code of Practice for In-service Inspection and Testing of Electrical Equipment* where appropriate. Where references to sections are made and extracts given, these may be found in the Code of Practice publication.

Outcome 2.1 Equipment construction

1 **A microwave oven has a mass of 21 kg and is located on a kitchen worktop. The equipment type is**

○ a   portable
○ b   transportable
○ c   built in
◉ d   stationary.

Answer d
See Section 5 (Types of electrical equipment) for a definition of stationary equipment or appliances, as well as definitions of portable, transportable and built-in equipment.

From Section 5.4 An item of stationary equipment or a stationary appliance is equipment that has a mass exceeding 18 kg and is not provided with a carrying handle, eg refrigerator, washing machine or dishwasher.

2 **A metal-cased electric iron has basic insulation around live parts and is supplied by a three-core cord. The appliance construction is Class**

◉ a   I
○ b   II
○ c   III
○ d   01.

Answer a
The question states that the supply is via a three-core cord, implying the need for a circuit protective conductor in addition to the live and neutral conductors, so the answer must be Class I. See Section 2 (Definitions) for definitions of Class I, Class II and Class III equipment. Class 0 and 01 are referred to in Section 11.4.

Notes

From Section 2 Class I equipment. Equipment in which protection against electric shock does not rely on basic insulation only, but which includes means for the connection of exposed-conductive-parts to a protective conductor in the fixed wiring of the installation.

3 **Which <u>one</u> of the following classification marks indicates an item of equipment that does not rely on installation conditions for protection against electric shock?**

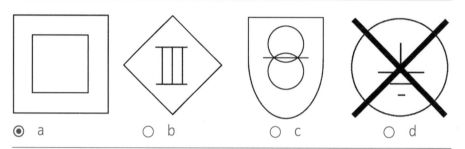

◉ a   ◯ b   ◯ c   ◯ d

Answer a

Option a shows the Class II construction mark. Class II equipment does not rely on installation conditions for protection against electric shock (see Section 11.2, Class II).

Option b shows the Class III mark and option c shows the safety isolating transformer mark.

From Section 2 Class II equipment. Equipment in which protection against electric shock does not rely on basic insulation only, but in which additional safety precautions such as supplementary insulation are provided, there being no provision for the connection of exposed metalwork of the equipment to a protective conductor and no reliance upon precautions to be taken in the fixed wiring of the installation.

4 **An electric shock through a failure of basic protection occurs when a person touches**

◯ a   the metal casing of a Class I appliance made live by a fault
◯ b   the earth terminal of an appliance in operation
◉ c   the phase terminal of an appliance in operation, which has its outer covers removed
◯ d   an exposed-conductive-part made live by a fault.

Answer c

An electric shock will occur when a live part is touched. Under normal operating conditions, all live parts are hidden either by insulation or by an enclosure. Should the insulation/enclosure be damaged or the outer covers be removed so that the live conductor is exposed, touching the live conductor will result in electric shock.

5   **In order for the appliance shown in the diagram below to be suitably protected in the event of a fault, the metal casing of the appliance must be**

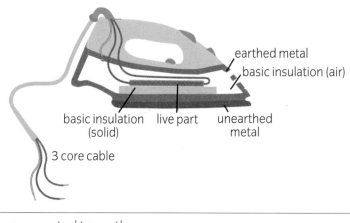

earthed metal
basic insulation (air)
basic insulation (solid)   live part   unearthed metal
3 core cable

From the *IEE Code of Practice for In-service Inspection and Testing of Electrical Equipment*, Section 11.1.

⦿ a   connected to earth
○ b   disconnected from earth
○ c   reinforced
○ d   non-ferrous.

Answer a

An item of Class I equipment must be provided with a protective earthing conductor in its power supply cable, for example a three-core flexible cord. See Section 11.1 (Class I).

From Section 11.1   Class I equipment includes appliances and tools, and for such equipment, protection against electric shock is provided by both
*   the provision of basic insulation, and
*   connecting metal parts to the protective conductor in the connecting cable and plug and hence via the socket-outlet to the fixed installation wiring and the means of earthing.

6   **If an appliance cord set length is doubled, the resistance value of the earth conductor will**

○ a   remain unchanged
○ b   decrease slightly
○ c   halve
◉ d   double.

Answer d

Table VII.1 (Nominal resistances of appliance supply cable protective conductors) in Appendix VII shows the nominal resistance of the protective conductor per metre length and for various lengths of cable that may be fitted as supply leads to appliances. It can be seen that doubling the length will double the resistance of the conductor.

Outcome 2.2   Inspection

7   **Portable equipment used in an office should be subjected to initial user checks**

○ a   before use
○ b   daily
◉ c   weekly
○ d   monthly.

Answer c

Portable equipment used in an office should be subjected to initial user checks on a weekly basis. See Table 7.1 in Section 7.

Table 7.1 Initial frequency of inspection and testing of equipment

	Equipment Use	Type of Equipment	User Checks	Class I		Class II	
			Not recorded unless a fault is found	Formal visual inspection Recorded	Combined inspection and testing Recorded	Formal visual inspection Recorded	Combined inspection and testing Recorded
	a	b	c	d	e	f	g
6	Offices and shops	S	none	24 months	48 months	24 months	None
		IT	none	24 months	48 months	24 months	None
		M	weekly	12 months	24 months	24 months	None
		P	**weekly**	12 months	24 months	24 months	None
		H	before use	6 months	12 months	6 months	None

From the *IEE Code of Practice for In-service Inspection and Testing of Electrical Equipment*, Section 7, extract from Table 7.1.

8 Which classification and location of equipment from the following is considered to be of the lowest risk?

- ○ a Class I equipment in an office
- ○ b Class I equipment in a factory
- ◉ c Class II equipment in an office
- ○ d Class II equipment on a construction site

Answer c

An office environment is considered to be less onerous than a factory or construction site and a Class II appliance does not rely on the integrity of the electrical installation, which makes it of the lowest risk. See Section 2 (Definitions) for a definition of Class II equipment.

Examiner's tip: Where a question asks for a comparison of risk between equipment or locations, Table 7.1 in Section 7 can prove useful. By looking at the table you can see that equipment or locations that require less frequent inspecting or testing are considered a lesser risk.

From Section 2 Class II equipment. Equipment in which protection against electric shock does not rely on basic insulation only, but in which additional safety precautions such as supplementary insulation are provided, there being no provision for the connection of exposed metalwork of the equipment to a protective conductor and no reliance upon precautions to be taken in the fixed wiring of the installation.

9 Initial formal visual inspection on hand-held Class I equipment used by the public should be carried out

- ◉ a   weekly
- ○ b   monthly
- ○ c   every 12 months
- ○ d   every 24 months.

Answer a

The initial formal visual inspection on hand-held Class I equipment used by the public should be carried out on a weekly basis. See Table 7.1 (Initial frequency of inspection and testing of equipment).

Table 7.1   Initial frequency of inspection and testing of equipment

	Equipment Use	Type of Equipment	User Checks	Class I		Class II	
				Formal visual inspection	Combined inspection and testing	Formal visual inspection	Combined inspection and testing
			Not recorded unless a fault is found	Recorded	Recorded	Recorded	Recorded
	a	b	c	d	e	f	g
3	Equipment used by the public	S	notes 3, 4	monthly	12 months	3 months	12 months
		IT	notes 3, 4	monthly	12 months	3 months	12 months
		M	notes 3, 4	weekly	6 months	1 month	12 months
		P	notes 3, 4	weekly	6 months	1 month	12 months
		H	notes 3, 4	**weekly**	6 months	1 month	12 months

From the *IEE Code of Practice for In-service Inspection and Testing of Electrical Equipment*, Section 7, extract from Table 7.1.

10 An earth continuity test on an item of Class I equipment produced a result of 0.05 Ω. The same test on the same item of equipment was carried out six months later and produced a result of 0.46 Ω. This would indicate that the equipment has

- ○ a   good insulation resistance
- ○ b   a better earth connection
- ○ c   a higher touch current
- ◉ d   a poorer earth connection.

Answer d

Ideally, the resistance of a protective conductor should be as low as possible. If the original reading was 0.05 Ω but on being re-tested a result of 0.46 Ω was obtained, this would indicate a significant deterioration in continuity and therefore a poorer earth connection would be achieved.

11 The <u>most</u> appropriate class of equipment for use outdoors would be Class

- ○ a   I
- ◉ b   II
- ○ c   0
- ○ d   01.

Answer b

Equipment is required to be suitable for the environment in which it is located (see Section 13, The user check). Unlike the other options, a Class II appliance does not rely on the integrity of the electrical installation (see Section 2, Definitions) and so would be most appropriate for outdoor use.

12 Any equipment showing a significant decrease in insulation resistance values, which remain within the allowed values, should be tested

- ◉ a   more frequently
- ○ b   less frequently
- ○ c   by the manufacturer
- ○ d   for dielectric strength.

Answer a

As long as the value of insulation resistance obtained is above the minimum acceptable values stated in Table 15.2 (Insulation resistance readings) in Section 15 the item may stay in service. However, a decrease in insulation resistance values obtained during testing is a sign of deterioration. As such it will be necessary to monitor the item of equipment more closely in order to assess its continued suitability for use. Carrying out inspection/testing more frequently would be an effective means of doing this.

Outcome 2.3 Combined inspection and testing

13 An earth continuity test on a 22 m extension lead having a core area of 2.5 mm² would produce a result of approximately

- ○ a   0.017 Ω
- ◉ b   0.176 Ω
- ○ c   0.821 Ω
- ○ d   8.00 Ω.

Answer b

Table VII.1 (Nominal resistances of appliance supply cable protective conductors) in Appendix VII states that the resistance of a 2.5 mm² conductor is 8 milli-ohm per metre (mΩ/m), which may also be written as 0.008 Ω/m.

The length of the cable given in the question is 22 m, therefore:

Resistance of protective conductor = 22 x 0.008
= 0.176 Ω

14 When conducting a formal visual inspection, care must be taken to ensure business equipment is isolated from normal electrical supply and any UPS connected. UPS is an abbreviation of

- ○ a   universal peripheral system
- ○ b   undervoltage protection system
- ○ c   universal provider supply
- ◉ d   uninterruptible power supply.

Answer d

UPS stands for uninterruptible power supply.

Business equipment may need to be powered down before isolation (see Section 14, The formal visual inspection) and specific procedures should be followed to prevent damage or loss of data, etc. Equipment supplied via an uninterruptible power supply (or other standby supply) must be isolated from its standby source before the inspection commences.

Notes

15 A detachable appliance cord set for Class I equipment must

- ○ a never have an earth core fitted
- ○ b always be protected by a 13 A fuse
- ○ c never have a non-rewireable type plug
- ◉ d always have a protective conductor connected.

Answer d

The cord set being used to supply an item of Class I equipment must always contain an effective protective conductor. See Section 11.1 (Class I).

From Section 11.1 Class I equipment relies for its safety upon a satisfactory means of earthing for the fixed installation and an adequate connection to it, normally via the flexible cable connecting the equipment, the plug and socket-outlet, and the circuit protective conductors of the fixed installation. Where Class I equipment is intended to be used with a flexible cable, the cable is required to include a protective conductor.

16 Where a Class I item of equipment is <u>not</u> fitted with a plug, earth continuity testing may be easier using

- ○ a a portable appliance test set with a 13 A socket
- ◉ b a continuity/insulation tester
- ○ c an earth loop impedance tester
- ○ d an RCD tester.

Answer b

Equipment can most easily be tested with a continuity/insulation tester. See Section 15.2 (Test procedures). However, earth continuity testing can also be carried out by a low resistance ohmmeter (see Section 10.3, Low resistance ohmmeters (for earth continuity testing)).

In the case of items of equipment connected via a switched, fused connection unit, no plug is fitted and so connection to the spring clips of a low resistance ohmmeter may be more convenient than using a portable appliance tester (option a), which in most cases is best suited for testing items of equipment with some form of plug fitted.

From Section 15.2 Equipment that is permanently connected to a flex outlet type of accessory can more easily be tested using an insulation/continuity test instrument with the test leads connected directly to the accessory terminals. The supply to the accessory is required to be isolated and proved dead at the point of work before the testing commences.

17 An insulation resistance test on a Class I appliance gives a test result of 0.02 MΩ. If the appliance is connected to a 230 V supply, the value of current that would leak to earth would be

○ a   1 mA
◉ b   11.5 mA
○ c   100 mA
○ d   1 A.

Answer b

The answer is found by using Ohm's Law, where V = IR.

Transpose the formula to make I the subject, ie

I = V/R
I = 230/0.02 MΩ
I = 11.5 mA

Examiner's tip: Remember that 0.02 MΩ is equivalent to 20,000 Ω. If using a scientific calculator then it may be calculated by:
230 ÷ 0.02 EXP 6 = as 0.02 MΩ is the same as multiplying the value by 10 six times.

The answer will probably be given as 0.0115. If you press the ENG button once, the value will be given in milli Amperes, which is the same as the display showing 11.5^{-3}.

Alternatively, using a basic calculator:
I = 230 ÷ 20000 = 0.0115 A = 11.5 mA.

18 If a touch current measurement is conducted as an alternative to insulation resistance testing, the test voltage used, for practical purposes, is

○ a   half the supply voltage
○ b   500 V a.c.
○ c   500 V d.c.
◉ d   the supply voltage.

Answer d

For practical purposes, the test voltage is the supply voltage. See Section 15.6 (Protective conductor/touch current measurement).

Outcome 2.4 Use of instruments, etc

19 A test for polarity is only required to be carried out on extension leads and

○ a   Class I hand-held appliances
○ b   non-detachable flexible cords
◉ c   detachable flexible cords
○ d   Class II hand-held appliances.

Answer c

Both extension leads and detachable flexible cords should be tested as Class I appliances, which includes a test for polarity. See Section 15.10.1 (Extension leads) and Section 15.9 (Appliance cord sets).

A polarity test is not specifically required for Class I and Class II appliances (options a and d), or for non-detachable flexible cords (option b).

20 Some electrical test equipment incorporates a test facility that is not recommended for in-service testing. This test facility is commonly known as

○ a   low pot testing
◉ b   flash testing
○ c   flush testing
○ d   no pot testing.

Answer b

See Section 15.3 (In-service tests). The other options are not correct because there are no tests specifically referred to as low pot, flush, or no pot within the *IEE Code of Practice*.

From Section 15.3 … Some electrical test devices apply tests which are inappropriate and may even damage equipment containing electronic circuits, possibly causing degradation to safety. In particular, whilst this Code of Practice includes insulation resistance tests, equipment should not be subjected to dielectric strength testing (also known as flash testing or hi-pot testing) because this may damage insulation and may also indirectly damage low voltage electronic circuits unless appropriate precautions are taken.

21 Which of the following test equipment would be required to carry out in-service inspection and testing?

- ○ a   Earth loop impedance tester
- ◉ b   Earth continuity tester
- ○ c   Prospective fault current tester
- ○ d   Clamp type ammeter

Answer b

In-service testing involves earth continuity tests, as well as insulation resistance testing and functional checks (see Section 15.3, In-service tests and Section 15.4, The earth continuity test). Equipment connected other than by plug and socket may be more easily tested by the use of a continuity/insulation tester (see Section 15.2, Test procedures).

Options a and c, the earth loop impedance tester and the prospective fault current tester, are used when carrying out tests on the fixed electrical installation whilst option d, the clamp type ammeter, is used to measure current flowing in a conductor without requiring disconnection.

22 An appliance is supplied with a 0.75 mm² flexible cord having a length of 2 m and an earth continuity test produces a reading of 0.172 Ω. The value of earth continuity for the appliance alone is

- ○ a   0.052 Ω and acceptable
- ○ b   0.072 Ω and unacceptable
- ○ c   0.1 Ω and acceptable
- ◉ d   0.12 Ω and unacceptable.

Answer d

The resistance value of 0.172 Ω is that of the appliance and the cord set.

Section 15.4 (The earth continuity test) states that the earth continuity resistance shall not exceed $(0.1 + R)$ Ω where R is the resistance of the protective conductor of the supply cord and 0.1 Ω for appliances without a supply cord. It is therefore necessary to find the resistance of the cord set, R, subtract that value from the total value; the result should be ≤ 0.1 Ω. Table VII.1 (Nominal resistances of appliance supply cable protective conductors) in Appendix VII shows that a 0.75 mm² flexible cord of length 2 m has a resistance value of 52 mΩ, ie 0.052 Ω. Now, subtract this value, R, from the total:

Therefore: $0.172 - 0.052 = 0.12$ Ω

It can be seen that 0.12 Ω is greater than 0.1 Ω, which is unacceptable.

23 Equipment labels should contain information such as

- ⦿ a date of check, safety status, appliance number and next test date
- ○ b safety status, appliance number, next test date and test results
- ○ c appliance number, next test date, test results and date of check
- ○ d next test date, test results, date of check and safety status.

Answer a

It is unnecessary to include the results of inspections/tests. These are recorded on the Equipment formal visual and combined inspection and test record (Form VI.2). Further, it should also be remembered that not all items (eg Class II equipment in certain environments) actually require formal testing to be carried out. Therefore, a is the only correct option. See Section 8.4 (Labelling).

From Section 8 The information provided should consist of an identification code to enable the equipment to be uniquely identifiable even if several similar items exist within the same premises. An indication of the current safety status of the equipment must also be included (eg whether the item has PASSed or FAILed the appropriate safety inspection/test). The date on which re-testing is due or the last test date and re-test period should also be stated.

From the *IEE Code of Practice for In-service Inspection and Testing of Electrical Equipment*, Appendix VI, Form VI.3.

24 Insulation deterioration may be occurring if previous insulation resistance test results recorded a reading that was, when compared to current results,

- ⦿ a   much higher
- ○ b   the same
- ○ c   slightly lower
- ○ d   much lower.

Answer a

In general terms, the resistance of insulation between live conductors and between live conductors and earth should be as high as possible. As insulation deteriorates, its resistance will fall. This is a clear indication that insulation has deteriorated or has been otherwise comprised.

Outcome 2.5   Equipment

25 Any equipment found to be unsafe following an inspection and test shall be

- ○ a   identified with a fail label and remain in use
- ○ b   identified with a pass label and withdrawn from use
- ⦿ c   identified with a fail label and withdrawn from use
- ○ d   identified with a pass label and remain in use.

Answer c

Equipment found to be unsafe must be identified with a fail label and removed from use. See Section 8.5 (Damaged or faulty equipment). It is also recommended that any items withdrawn from service should be left with the person responsible for managing the in-service inspection and testing of electrical equipment within the premises. This is in case any disputes arise over missing items.

From Section 8.5 If equipment is found to be damaged or faulty on inspection or test, an assessment should be made by a responsible person as to the suitability of the equipment for the use in that particular location. … If it is unsuitable it should be replaced by more suitable equipment.

26 An extension lead has a cross-sectional area (csa) of 1.5 mm^2 and a length of 14 m. An earth continuity reading should give a result of approximately

○ a   13.3 mΩ
○ b   19 mΩ
◉ c   0.186 Ω
○ d   186.2 Ω.

Answer c

Table VII.1 (Nominal resistances of appliance supply cable protective conductors) in Appendix VII shows that a 1.5 mm^2 flexible cord having a length of 1 m has a resistance value of 13.3 mΩ, ie 0.0133 Ω.

Since conductor resistance is proportional to its length, multiply this value by 14.

Therefore: 0.0133 x 14 = 0.1862 Ω

Rounded off to three decimal places, the answer is 0.186 Ω

Examiner's tip: Remember that the values given in Table VII.1 are in milli-ohms, therefore the answer you get from your calculator must be divided by 1000 to convert the answer to ohms, in other words, if you calculated the answer as 13.3 x 14 ÷ 1000 (= 0.186).

27 The purpose of the fuse in a standard 13 A plug is to

○ a   protect the appliance
○ b   protect the fixed installation
○ c   conform with European standards
◉ d   protect the flexible cord.

Answer d

The primary purpose of the fuse in the 13 A plug is to protect the flexible cord and to allow for a reduction in cross-sectional area of the conductors within the cord. See Section 15.14 (Plug fuses).

Circuits within the fixed installation are protected by overcurrent and earth leakage devices as appropriate to meet the requirements of *BS 7671: Requirements for Electrical Installations*.

From Section 15.14   The fuse in the plug is not fitted to protect the appliance, although in practice it often does this. The fuse in the plug protects the flex against faults and can allow the use of a reduced csa flexible cable. This is advantageous for appliances such as electric blankets, soldering irons and Christmas tree lights, where the flexibility of a small flexible cable is desirable.

28 **Equipment having a high protective conductor current exceeding 10 mA requires specific precautions to be observed. Further information on the arrangement of such equipment can be found in**

- ○ a   BS 7176
- ○ b   GS38
- ◉ c   BS 7671
- ○ d   PM28.

Answer c

There are particular requirements in *BS 7671: Requirements for Electrical Installations* for the earthing arrangements for equipment having high protective conductor currents. See Section 15.12 (High protective conductor currents).

29 **A warning label is required for any item of equipment having a protective conductor current in excess of**

- ○ a   2.5 mA
- ◉ b   3.5 mA
- ○ c   5 mA
- ○ d   6.5 mA.

Answer b

Equipment with a protective conductor current in excess of 3.5 mA should have a warning label fixed adjacent to the equipment primary power connection. See Section 15.12 (High protective conductor currents).

The Code of Practice states that the label should bear the following warning, or similar wording:

WARNING: HIGH PROTECTIVE CONDUCTOR CURRENT
Earth connection essential before connecting the supply

30 A detachable, three-core cord set used to supply 230 V information technology equipment shall be tested as Class

- ● a   I
- ○ b   II
- ○ c   III
- ○ d   0.

Answer a

Three-core cord sets should be tested as a Class I appliance with the cord set plugged into the appliance. See Section 15.9 (Appliance cord sets).

Two-core cord sets should be tested as Class II appliances. The *IEE Code of Practice* does not cover the in-service inspection and testing of Class 0 or Class III items of equipment.

Notes

Answer key

Sample test 2

Question	Answer
1	d
2	a
3	a
4	c
5	a
6	d
7	c
8	c
9	a
10	d
11	b
12	a
13	b
14	d
15	d
16	b
17	b
18	d
19	c
20	b
21	b
22	d
23	a
24	a
25	c
26	c
27	d
28	c
29	b
30	a

More information

Further reading 114

Online resources 115

Further courses 116

More information

Further reading

Required reading
IEE Code of Practice for In-service Inspection and Testing of Electrical Equipment, Third Edition, published by the IEE, London, 2008

Additional reading
BS 7671: 2008 Requirements for Electrical Installations, IEE Wiring Regulations Seventeenth Edition, published by the IEE, London, 2008

IEE Guidance Notes, a series of guidance notes, each of which enlarges upon and amplifies the particular requirements of a part of the Wiring Regulations, published by the IEE, London:
– Guidance Note 3: *Inspection and Testing*, 5th edition 2008

The Electrician's Guide to Good Electrical Practice, published by Amicus, 2005

Electrician's Guide to the Building Regulations, published by the IEE, London, 2005

Brian Scaddan, *Electrical Installation Work*, published by Newnes (an imprint of Butterworth-Heinemann), 2002

John Whitfield, *Electrical Craft Principles*, published by the IEE, London, 1995

Online resources

City & Guilds www.cityandguilds.com
The City & Guilds website can give you more information about studying for further professional and vocational qualifications to advance your personal or career development, as well as locations of centres that provide the courses.

Institution of Engineering and Technology (IET) www.theiet.org
The Institution of Engineering and Technology was formed by the amalgamation of the Institution of Electrical Engineers (IEE) and the Institution of Incorporated Engineers (IIE). It is the largest professional engineering society in Europe and the second largest of its type in the world. The Institution produces the *IEE Wiring Regulations* and a range of supporting material and courses.

SmartScreen www.smartscreen.co.uk
City & Guilds' dedicated online support portal SmartScreen provides learner and tutor support for over 100 City & Guilds qualifications. It helps engage learners in the excitement of learning and enables tutors to free up more time to do what they love the most – teach!

BRE Certification Ltd www.partp.co.uk

British Standards Institution www.bsi-global.com

CORGI Services Ltd www.corgi-gas-safety.com

ELECSA Ltd www.elecsa.org.uk

Electrical Contractors' Association (ECA) www.eca.co.uk

Joint Industry Board for the Electrical Contracting Industry (JIB)
www.jib.org.uk

NAPIT Certification Services Ltd www.napit.org.uk

National Inspection Council for Electrical Installation Contracting (NICEIC) www.niceic.org.uk

Oil Firing Technical Association for the Petroleum Industry (OFTEC)
www.oftec.co.uk

Further courses

City & Guilds Level 3 Certificate in the Requirements for Electrical Installations BS 7671: 2008 (2382-10)

This qualification is aimed at practising electricians with relevant experience and is intended to ensure that they are conversant with the format, content and application of BS 7671: Requirements for Electrical Installations, 17th Edition.

City & Guilds Level 3 Certificate in Inspection, Testing and Certification of Electrical Installations (2391-10)

This course is aimed at those with practical experience of inspection and testing of LV electrical installations, who require to become certificated possibly for NICEIC purposes. It is not suitable for beginners. In addition to relevant practical experience, candidates must possess a good working knowledge of the requirements of BS 7671 to City & Guilds Level 3 certificate standard or equivalent.

City & Guilds Level 2 Certificate in Fundamental Inspection, Testing and Initial Verification (2392-10)

This qualification provides candidates with an introduction to the initial verification of electrical installations. It is aimed at practising electricians who have not carried out inspection and testing since qualifying, those who require update training and those with limited experience of inspection and testing. Together with suitable on-site experience, it also prepares candidates to go on to the Level 3 Certificate in Inspection, Testing and Certification of Electrical Installations (2391-10).

City & Guilds Building Regulations for Electrical Safety

This new suite of qualifications is for Competent Persons in Domestic Electrical Installations (Part P of the Building Regulations). The qualifications consist of components for specialised domestic building regulations and domestic wiring regulations routes as well as a component for Qualified Supervisors.

JIB Electrotechnical Certification Scheme (ECS) Health and Safety Assessment

This Health and Safety Assessment is a requirement for electricians wishing to work on larger construction projects and sites in the UK and the exam is an online type very similar in format to the GOLA tests. It is now a mandatory requirement for holding an ECS card, and is a requirement for all members of the ECS. Please refer to www.jib.org.uk/ecs2.htm for details.